Numerical Recipes
Example Book (FORTRAN)

Second Edition

Numerical Recipes
Example Book (FORTRAN)

Second Edition

William T. Vetterling

Polaroid Corporation

Saul A. Teukolsky

Department of Physics, Cornell University

William H. Press

Harvard-Smithsonian Center for Astrophysics

Brian P. Flannery

EXXON Research and Engineering Company

CAMBRIDGE
UNIVERSITY PRESS

Published by the Press Syndicate of the University of Cambridge
The Pitt Building, Trumpington Street, Cambridge CB2 1RP
40 West 20th Street, New York, NY 10011-4211, USA
10 Stamford Road, Oakleigh, Melbourne 3166, Australia

First edition originally published 1986
Second edition originally published 1992
Reprinted 1993 (twice), 1994

Printed in the United States of America
Typeset in TEX

The computer programs in this book are available, in
FORTRAN, in several machine-readable formats. There are
also versions of this book and its software available in C,
Pascal, and BASIC programming languages.

To purchase diskettes in IBM PC or Apple Macintosh
formats, use the order form at the back of the book or write
to Cambridge University Press, 110 Midland Avenue, Port
Chester, NY 10573.

Unlicenced transfer of Numerical Recipes programs from
the above-mentioned IBM PC or Apple Macintosh
diskettes to any other format or to any computer except a
single IBM PC or Apple Macintosh or compatible for each
diskette purchased, is strictly prohibited. Licenses for
authorized transfers to other computers are available from
Numerical Recipes Software, P.O. Box 243, Cambridge,
MA 12238 (FAX 617 863-1739). Technical questions,
corrections, and requests for information on other available
formats should be directed to this address.

Library of Congress Cataloging-in-Publication Data available.

A catalogue record for this book is available from the British Library.

ISBN 0-521-43721-0 Example book in FORTRAN (this book)
ISBN 0-521-43064-X Numerical Recipes in FORTRAN
ISBN 0-521-43717-2 FORTRAN diskette (IBM 5.25", 1.2M)
ISBN 0-521-43719-9 FORTRAN diskette (IBM 3.5", 720K)
ISBN 0-521-43716-4 FORTRAN diskette (Mac 3.5". 800K)

CONTENTS

Preface

This *Numerical Recipes Example Book (FORTRAN)* is designed to accompany the text and reference book *Numerical Recipes in FORTRAN: The Art of Scientific Computing*, Second Edition, by William H. Press, Saul A. Teukolsky, William T. Vetterling, and Brian P. Flannery (Cambridge University Press, 1992). In that volume, the algorithms and methods of scientific computation are developed in considerable detail, starting with basic mathematical analysis and working through to actual implementation in the form of FORTRAN subroutines. The routines in *Numerical Recipes in FORTRAN: The Art of Scientific Computing*, numbering more than 300, are meant to be incorporated into user applications; they are subroutines (or functions), not stand-alone programs.

It often happens, when you want to incorporate somebody else's procedure into your own application program, that you first want to see the procedure demonstrated on a simple example. Prose descriptions of how to use a procedure (even those in *Numerical Recipes*) can occasionally be inexact. There is no substitute for an actual, FORTRAN demonstration program that shows exactly how data are fed to a procedure, how the procedure is called, and how its results are unloaded and interpreted.

Another not unusual case occurs when you have, for one seemingly good purpose or another, modified the source code in a "foreign" procedure. In such circumstances, you might well want to test the modified procedure on an example known previously to have worked correctly, *before* letting it loose on your own data. There is the related case where procedure source code may have become corrupted, e.g., lost some lines or characters in transmission from one machine to another, and a simple revalidation test is desirable.

These are the needs addressed by this *Numerical Recipes Example Book*. Divided into chapters identically with *Numerical Recipes in FORTRAN: The Art of Scientific Computing*, this book contains FORTRAN source programs that exercise and demonstrate all of the *Numerical Recipes* subroutines and functions. The programs are commented, and each is also prefaced by a short description of what it does, and of which *Numerical Recipes* routines it exercises. In many cases where the demonstration programs require input data, that data is also printed in this book. In some cases, where the demonstration programs are not "self-validating," sample output is also shown.

Necessarily, in the interests of clarity, the *Numerical Recipes* procedures and functions are demonstrated in simple ways. A consequence is that the demonstration programs in this book do not usually test all possible regimes of input data, or even all lines of procedure source code. The demonstration programs in this book were by no means the only validating tests that the *Numerical Recipes* procedures and functions were required to pass during their development. The programs in this book *were* used during the later stages of the production of *Numerical Recipes in FORTRAN: The Art of Scientific Computing* to maintain integrity of the source code, and in this role were found to be invaluable.

DISCLAIMER OF WARRANTY

THE PROGRAMS LISTED IN THIS BOOK ARE PROVIDED "AS IS" WITHOUT WARRANTY OF ANY KIND. WE MAKE NO WARRANTIES, EXPRESS OR IMPLIED, THAT THE PROGRAMS CONTAINED IN THIS VOLUME ARE FREE OF ERROR, OR ARE CONSISTENT WITH ANY PARTICULAR STANDARD OF MERCHANTABILITY, OR THAT THEY WILL MEET YOUR REQUIREMENTS FOR ANY PARTICULAR APPLICATION. THEY SHOULD NOT BE RELIED ON FOR SOLVING A PROBLEM WHOSE INCORRECT SOLUTION COULD RESULT IN INJURY TO A PERSON OR LOSS OF PROPERTY. IF YOU DO USE THE PROGRAMS IN SUCH A MANNER, IT IS AT YOUR OWN RISK. THE AUTHORS AND PUBLISHER DISCLAIM ALL LIABILITY FOR DIRECT OR CONSEQUENTIAL DAMAGES RESULTING FROM YOUR USE OF THE PROGRAMS.

Chapter 1: Preliminaries

The routines in Chapter 1 of Numerical Recipes are introductory and less general in purpose than those in the remainder of the book. This chapter's routines serve primarily to expose the book's notational conventions, illustrate control structures, and perhaps to amuse. You may even find them useful. We hope that you will use badluk *for no serious purpose.*

⋆   ⋆   ⋆   ⋆

Subroutine flmoon calculates the phases of the moon, or more exactly, the Julian day and fraction thereof on which a given phase will occur or has occurred. The program xflmoon asks the present date and compiles a list of upcoming phases. We have compared the predictions to lunar tables, with happy results. Shown are the results of a test run, which you may replicate as a check. In this program, notice that we have set TZONE (the time zone) to −5.0 to signify the five hour separation of the Eastern Standard time zone from Greenwich, England. Our convention requires you to use negative values of TZONE if you are west of Greenwich, as we are. The Julian day results are converted to calendar dates through the use of caldat, which appears later in the chapter. The fractional Julian day and time zone combine to form a correction that can possibly change the calendar date by one day.

```
    Date        Time(EST)    Phase
12  9 1992       7 PM       full moon
12 16 1992       2 PM       last quarter
12 23 1992       8 PM       new moon
12 31 1992      10 PM       first quarter
 1  8 1993       8 AM       full moon
 1 14 1993      11 PM       last quarter
 1 22 1993       1 PM       new moon
 1 30 1993       6 PM       first quarter
 2  6 1993       7 PM       full moon
 2 13 1993      10 AM       last quarter
 2 21 1993       8 AM       new moon
 3  1 1993      11 AM       first quarter
 3  8 1993       5 AM       full moon
 3 14 1993      11 PM       last quarter
 3 23 1993       2 AM       new moon
 3 30 1993      11 PM       first quarter
 4  6 1993       2 PM       full moon
 4 13 1993       3 PM       last quarter
 4 21 1993       6 PM       new moon
 4 29 1993       8 AM       first quarter
```

```
      PROGRAM xflmoon
C     driver for routine flmoon
      REAL TZONE
      PARAMETER(TZONE=-5.0)
      REAL frac,timzon
      INTEGER i,im,id,iy,ifrac,istr,j1,j2,julday,n,nph
      CHARACTER phase(4)*15,timstr(2)*3
      DATA phase/'new moon','first quarter',
     *     'full moon','last quarter'/
      DATA timstr/' AM',' PM'/
      write(*,*) 'Date of the next few phases of the moon'
      write(*,*) 'Enter today''s date (e.g. 12,15,1992)'
      timzon=TZONE/24.0
      read(*,*) im,id,iy
C     approximate number of full moons since January 1900
      n=12.37*(iy-1900+(im-0.5)/12.0)
      nph=2
      j1=julday(im,id,iy)
      call flmoon(n,nph,j2,frac)
      n=n+nint((j1-j2)/29.53)
      write(*,'(/1x,t6,a,t19,a,t32,a)') 'Date','Time(EST)','Phase'
      do 11 i=1,20
        call flmoon(n,nph,j2,frac)
        ifrac=nint(24.*(frac+timzon))
        if (ifrac.lt.0) then
          j2=j2-1
          ifrac=ifrac+24
        endif
        if (ifrac.ge.12) then
          j2=j2+1
          ifrac=ifrac-12
        else
          ifrac=ifrac+12
        endif
        if (ifrac.gt.12) then
          ifrac=ifrac-12
          istr=2
        else
          istr=1
        endif
        call caldat(j2,im,id,iy)
        write(*,'(1x,2i3,i5,t20,i2,a,5x,a)') im,id,iy,
     *        ifrac,timstr(istr),phase(nph+1)
        if (nph.eq.3) then
          nph=0
          n=n+1
        else
          nph=nph+1
        endif
11    continue
      END
```

The function `julday`, our exemplar of the `if` control structure, converts calendar dates to Julian dates. Not many people know the Julian date of their birthday or any other convenient reference point, for that matter. To remedy this, we offer a list of checkpoints, which appears at the end of this chapter as the file `DATES.DAT`. The program xjulday lists the Julian date for each historic event for comparison. Then it allows you to make your own choices for entertainment.

```
      PROGRAM xjulday
C     driver for julday
      INTEGER i,im,id,iy,julday,n
      CHARACTER txt*40,name(12)*15
      DATA name/'January','February','March','April','May','June',
     *    'July','August','September','October','November',
     *    'December'/
      open(7,file='DATES.DAT',status='OLD')
      read(7,'(a)') txt
      read(7,*) n
      write(*,'(/1x,a,t12,a,t17,a,t23,a,t37,a/)') 'Month','Day','Year',
     *    'Julian Day','Event'
      do 11 i=1,n
        read(7,'(i2,i3,i5,a)') im,id,iy,txt
        write(*,'(1x,a10,i3,i6,3x,i7,5x,a)') name(im),id,iy,
     *      julday(im,id,iy),txt
11    continue
      close(7)
      write(*,'(/1x,a/)') 'Month,Day,Year (e.g. 1,13,1905)'
      do 12 i=1,20
        write(*,*) 'MM,DD,YYYY'
        read(*,*) im,id,iy
        if (im.lt.0) stop
        write(*,'(1x,a12,i8/)') 'Julian Day: ',julday(im,id,iy)
12    continue
      END
```

The next program in *Numerical Recipes* is badluk, an infamous code that combines the best and worst instincts of man. We include no demonstration program for badluk, not just because we fear it, but also because it is self-contained, with sample results appearing in the text.

Chapter 1 closes with routine caldat, which illustrates no new points, but complements julday by doing conversions from Julian day number to the month, day, and year on which the given Julian day began. This offers an opportunity, grasped by the demonstration program xcaldat, to push dates through both julday and caldat in succession, to see if they survive intact. This, of course, tests only your authors' ability to make mistakes backwards as well as forwards, but we hope you will share our optimism that correct results here speak well for both routines. (We have checked them a bit more carefully in other ways.)

```
      PROGRAM xcaldat
C     driver for routine caldat
      INTEGER i,im,imm,id,idd,iy,iyy,iycopy,j,julday,n
      CHARACTER name(12)*10
C     check whether CALDAT properly undoes the operation of JULDAY
      DATA name/'January','February','March','April','May',
     *    'June','July','August','September','October',
     *    'November','December'/
      open(7,file='DATES.DAT',status='OLD')
      read(7,*)
      read(7,*) n
      write(*,'(/1x,a,t40,a)') 'Original Date:','Reconstructed Date:'
```

```
      write(*,'(1x,a,t12,a,t17,a,t25,a,t40,a,t50,a,t55,a/)')
     *        'Month','Day','Year','Julian Day','Month','Day','Year'
      do 11 i=1,n
        read(7,'(i2,i3,i5)') im,id,iy
        iycopy=iy
        j=julday(im,id,iycopy)
        call caldat(j,imm,idd,iyy)
        write(*,'(1x,a,i3,i6,4x,i9,6x,a,i3,i6)') name(im),id,
     *        iy,j,name(imm),idd,iyy
11    continue
      END
```

Appendix

File DATES.DAT:

```
List of dates for testing routines in Chapter 1
16 entries
12 31   -1 End of millennium
01 01    1 One day later
10 14 1582 Day before Gregorian calendar
10 15 1582 Gregorian calendar adopted
01 17 1706 Benjamin Franklin born
04 14 1865 Abraham Lincoln shot
04 18 1906 San Francisco earthquake
05 07 1915 Sinking of the Lusitania
07 20 1923 Pancho Villa assassinated
05 23 1934 Bonnie and Clyde eliminated
07 22 1934 John Dillinger shot
04 03 1936 Bruno Hauptman electrocuted
05 06 1937 Hindenburg disaster
07 26 1956 Sinking of the Andrea Doria
06 05 1976 Teton dam collapse
05 23 1968 Julian Day 2440000
```

Chapter 2: Linear Algebraic Equations

Numerical Recipes Chapter 2 begins the "true grit" of numerical analysis by considering the solution of linear algebraic equations. This is done first by Gauss-Jordan elimination (gaussj), and then by LU decomposition with forward and backsubstitution (ludcmp and lubksb). Several linear systems of special form, represented by tridiagonal, band diagonal, cyclic, Vandermonde, and Toeplitz matrices, may be treated with subroutines tridag, bandec, cyclic, vander, and toeplz respectively. Cholesky decomposition (choldc and cholsl) is the preferred method for symmetric positive definite systems. QR decomposition (qrdcmp) is less efficient than LU decomposition in general; however, the ease with which it is updated (qrupdt) if the system is changed slightly makes it useful in certain applications. For singular or nearly singular matrices the best choice is singular value decomposition with backsubstitution (svdcmp and svbksb). Linear systems with relatively few non-zero coefficients, so-called "sparse" matrices, are handled by routine linbcg. A suite of routines for manipulating general sparse matrices, sprsax, sprstx, etc., is provided.

★ ★ ★ ★

gaussj performs Gauss-Jordan elimination with full pivoting to find the solution of a set of linear equations for a collection of right-hand side vectors. The demonstration routine xgaussj checks its operation with reference to a group of test input matrices printed at the end of this chapter as file MATRX1.DAT. Each matrix is subjected to inversion by gaussj, and then multiplication by its own inverse to see that a unit matrix is produced. Then the solution vectors are each checked through multiplication by the original matrix and comparison with the right-hand side vectors that produced them.

```
      PROGRAM xgaussj
C     driver for routine gaussj
      INTEGER MP,NP
      PARAMETER(MP=20,NP=20)
      INTEGER j,k,l,m,n
      REAL a(NP,NP),b(NP,MP),ai(NP,NP),x(NP,MP)
      REAL u(NP,NP),t(NP,MP)
      CHARACTER dummy*3
      open(7,file='MATRX1.DAT',status='old')
10    read(7,'(a)') dummy
      if (dummy.eq.'END') goto 99
      read(7,*)
      read(7,*) n,m
      read(7,*)
```

```
      read(7,*) ((a(k,l), l=1,n), k=1,n)
      read(7,*)
      read(7,*) ((b(k,l), k=1,n), l=1,m)
C     save Matrices for later testing of results
      do 13 l=1,n
        do 11 k=1,n
          ai(k,l)=a(k,l)
11      continue
        do 12 k=1,m
          x(l,k)=b(l,k)
12      continue
13    continue
C     invert Matrix
      call gaussj(ai,n,NP,x,m,MP)
      write(*,*) 'Inverse of Matrix A : '
      do 14 k=1,n
        write(*,'(1h ,(6f12.6))') (ai(k,l), l=1,n)
14    continue
C     test Results
C     check Inverse
      write(*,*) 'A times A-inverse (compare with unit matrix)'
      do 17 k=1,n
        do 16 l=1,n
          u(k,l)=0.0
          do 15 j=1,n
            u(k,l)=u(k,l)+a(k,j)*ai(j,l)
15        continue
16      continue
        write(*,'(1h ,(6f12.6))') (u(k,l), l=1,n)
17    continue
C     check Vector Solutions
      write(*,*) 'Check the following vectors for equality:'
      write(*,'(t12,a8,t23,a12)') 'Original','Matrix*Sol''n'
      do 21 l=1,m
        write(*,'(1x,a,i2,a)') 'Vector ',l,':'
        do 19 k=1,n
          t(k,l)=0.0
          do 18 j=1,n
            t(k,l)=t(k,l)+a(k,j)*x(j,l)
18        continue
          write(*,'(8x,2f12.6)') b(k,l),t(k,l)
19      continue
21    continue
      write(*,*) '*********************************'
      write(*,*) 'Press RETURN for next problem:'
      read(*,*)
      goto 10
99    close(7)
      END
```

The demonstration program for routine ludcmp relies on the same package of test matrices, but just performs an LU decomposition of each. The performance is checked by multiplying the lower and upper matrices of the decomposition and comparing with the original matrix. The array indx keeps track of the scrambling done by ludcmp to effect partial pivoting. We had to do the unscrambling here, but you will normally not be called upon to do so, since ludcmp is used with the routine lubksb, which knows how to do its own descrambling.

```
      PROGRAM xludcmp
C     driver for routine ludcmp
      INTEGER NP
      PARAMETER(NP=20)
      INTEGER j,k,l,m,n,indx(NP),jndx(NP)
      REAL d,dum,a(NP,NP),xl(NP,NP),xu(NP,NP),x(NP,NP)
      CHARACTER txt*3
      open(7,file='MATRX1.DAT',status='old')
      read(7,*)
10    read(7,*)
      read(7,*) n,m
      read(7,*)
      read(7,*) ((a(k,l), l=1,n), k=1,n)
      read(7,*)
      read(7,*) ((x(k,l), k=1,n), l=1,m)
C     print out a-matrix for comparison with product of lower
C     and upper decomposition matrices.
      write(*,*) 'Original matrix:'
      do 11 k=1,n
        write(*,'(1x,6f12.6)') (a(k,l), l=1,n)
11    continue
C     perform the decomposition
      call ludcmp(a,n,NP,indx,d)
C     compose separately the lower and upper matrices
      do 13 k=1,n
        do 12 l=1,n
          if (l.gt.k) then
            xu(k,l)=a(k,l)
            xl(k,l)=0.0
          else if (l.lt.k) then
            xu(k,l)=0.0
            xl(k,l)=a(k,l)
          else
            xu(k,l)=a(k,l)
            xl(k,l)=1.0
          endif
12      continue
13    continue
C     compute product of lower and upper matrices for
C     comparison with original matrix.
      do 16 k=1,n
        jndx(k)=k
        do 15 l=1,n
          x(k,l)=0.0
          do 14 j=1,n
            x(k,l)=x(k,l)+xl(k,j)*xu(j,l)
14        continue
15      continue
16    continue
      write(*,*) 'Product of lower and upper matrices (unscrambled):'
      do 17 k=1,n
        dum=jndx(indx(k))
        jndx(indx(k))=jndx(k)
        jndx(k)=dum
17    continue
```

```
      do 19 k=1,n
        do 18 j=1,n
          if (jndx(j).eq.k) then
            write(*,'(1x,6f12.6)') (x(j,l), l=1,n)
          endif
18      continue
19    continue
      write(*,*) 'Lower matrix of the decomposition:'
      do 21 k=1,n
        write(*,'(1x,6f12.6)') (xl(k,l), l=1,n)
21    continue
      write(*,*) 'Upper matrix of the decomposition:'
      do 22 k=1,n
        write(*,'(1x,6f12.6)') (xu(k,l), l=1,n)
22    continue
      write(*,*) '*********************************'
      write(*,*) 'Press RETURN for next problem:'
      read(*,*)
      read(7,'(a3)') txt
      if (txt.ne.'END') goto 10
      close(7)
      END
```

Our example driver for `lubksb` makes calls to both `ludcmp` and `lubksb` in order to solve the linear equation problems posed in file `MATRX1.DAT` (see discussion of `gaussj`). The original matrix of coefficients is applied to the solution vectors to check that the result matches the right-hand side vectors posed for each problem. We apologize for using routine `ludcmp` in a test of `lubksb`, but `ludcmp` has been tested independently, and anyway, `lubksb` is nothing without this partner program, so a test of the combination is more to the point.

```
      PROGRAM xlubksb
C     driver for routine lubksb
      INTEGER NP
      PARAMETER (NP=20)
      REAL p,a(NP,NP),b(NP,NP),c(NP,NP),x(NP)
      INTEGER j,k,l,m,n,indx(NP)
      CHARACTER txt*3
      open(7,file='MATRX1.DAT',status='old')
      read(7,*)
10    read(7,*)
      read(7,*) n,m
      read(7,*)
      read(7,*) ((a(k,l), l=1,n), k=1,n)
      read(7,*)
      read(7,*) ((b(k,l), k=1,n), l=1,m)
C     save matrix a for later testing
      do 12 l=1,n
        do 11 k=1,n
          c(k,l)=a(k,l)
11      continue
12    continue
C     do LU decomposition
      call ludcmp(c,n,NP,indx,p)
C     solve equations for each right-hand vector
      do 16 k=1,m
```

```
      do 13 l=1,n
        x(l)=b(l,k)
13    continue
      call lubksb(c,n,NP,indx,x)
C     test results with original matrix
      write(*,*) 'Right-hand side vector:'
      write(*,'(1x,6f12.6)') (b(l,k), l=1,n)
      write(*,*) 'Result of matrix applied to sol''n vector'
      do 15 l=1,n
        b(l,k)=0.0
        do 14 j=1,n
          b(l,k)=b(l,k)+a(l,j)*x(j)
14      continue
15    continue
      write(*,'(1x,6f12.6)') (b(l,k), l=1,n)
      write(*,*) '*********************************'
16    continue
      write(*,*) 'Press RETURN for next problem:'
      read(*,*)
      read(7,'(a3)') txt
      if (txt.ne.'END') goto 10
      close(7)
      END
```

Subroutine `tridag` solves linear equations with coefficients that form a tridiagonal matrix. We provide at the end of this chapter a second file of matrices MATRIX2.DAT for the demonstration driver. In all other respects, the demonstration program `xtridag` operates in the same fashion as `xlubksb`.

```
      PROGRAM xtridag
C     driver for routine tridag
      INTEGER NP
      PARAMETER (NP=20)
      INTEGER k,n
      REAL diag(NP),superd(NP),subd(NP),rhs(NP),u(NP)
      CHARACTER txt*3
      open(7,file='MATRX2.DAT',status='old')
10    read(7,'(a3)') txt
      if (txt.eq.'END') goto 99
      read(7,*)
      read(7,*) n
      read(7,*)
      read(7,*) (diag(k), k=1,n)
      read(7,*)
      read(7,*) (superd(k), k=1,n-1)
      read(7,*)
      read(7,*) (subd(k), k=2,n)
      read(7,*)
      read(7,*) (rhs(k), k=1,n)
C     carry out solution
      call tridag(subd,diag,superd,rhs,u,n)
      write(*,*) 'The solution vector is:'
      write(*,'(1x,6f12.6)') (u(k), k=1,n)
C     test solution
      write(*,*) '(matrix)*(sol''n vector) should be:'
      write(*,'(1x,6f12.6)') (rhs(k), k=1,n)
      write(*,*) 'Actual result is:'
```

```
      do 11 k=1,n
        if (k.eq.1) then
          rhs(k)=diag(1)*u(1) + superd(1)*u(2)
        else if (k.eq.n) then
          rhs(k)=subd(n)*u(n-1) + diag(n)*u(n)
        else
          rhs(k)=subd(k)*u(k-1) + diag(k)*u(k)
     *            + superd(k)*u(k+1)
        endif
11    continue
      write(*,'(1x,6f12.6)') (rhs(k), k=1,n)
      write(*,*) '*********************************'
      write(*,*) 'Press RETURN for next problem:'
      read(*,*)
      goto 10
99    close(7)
      END
```

The demonstration program xbanmul forms a banded matrix, multiplies if by a vector using banmul, and compares the answer to the result of carrying out a full matrix multiplication.

```
      PROGRAM xbanmul
C     driver for routine banmul
      INTEGER NP,M1,M2,MP
      PARAMETER (NP=7,M1=2,M2=1,MP=M1+1+M2)
      INTEGER i,j,k
      REAL a(NP,MP),aa(NP,NP),ax(NP),b(NP),x(NP)
      do 12 i=1,M1
        do 11 j=1,NP
          a(j,i)=10*j+i
11      continue
12    continue
C     lower band
      do 13 i=1,NP
        a(i,M1+1)=i
13    continue
C     diagonal
      do 15 i=1,M2
        do 14 j=1,NP
          a(j,M1+1+i)=0.1*j+i
14      continue
15    continue
C     upper band
      do 17 i=1,NP
        do 16 j=1,NP
          k=i-M1-1
          if (j.ge.max(1,1+k).and.j.le.min(M1+M2+1+k,NP)) then
            aa(i,j)=a(i,j-k)
          else
            aa(i,j)=0.0
          endif
16      continue
17    continue
      do 18 i=1,NP
        x(i)=i/10.0
18    continue
```

```
      call banmul(a,NP,M1,M2,NP,MP,x,b)
      do 21 i=1,NP
        ax(i)=0.0
        do 19 j=1,NP
          ax(i)=ax(i)+aa(i,j)*x(j)
19      continue
21    continue
      write(*,'(t8,a,t32,a)') 'Reference vector','banmul vector'
      do 22 i=1,NP
        write(*,'(t8,f12.4,t32,f12.4)') ax(i),b(i)
22    continue
      END
```

The sample program xbandec forms a random banded matrix **A**, a random vector x, and calculates $b = A \cdot x$. It then supplies **A** and **b** to bandec and banbks and compares the solution to the saved copy of x.

```
      PROGRAM xbandec
C     driver for routine bandec
      REAL ran1
      REAL a(7,4),x(7),b(7),al(7,2),d
      INTEGER indx(7)
      INTEGER i,idum,j
      idum=-1
      do 12 i=1,7
        x(i)=ran1(idum)
        do 11 j=1,4
          a(i,j)=ran1(idum)
11      continue
12    continue
      call banmul(a,7,2,1,7,4,x,b)
      do 13 i=1,7
        write(*,*) i,b(i),x(i)
13    continue
      call bandec(a,7,2,1,7,4,al,2,indx,d)
      call banbks(a,7,2,1,7,4,al,2,indx,b)
      do 14 i=1,7
        write(*,*) i,b(i),x(i)
14    continue
      stop
      END
```

mprove is a short routine for improving the solution vector for a set of linear equations, providing that an LU decomposition has been performed on the matrix of coefficients. Our test of this function is to use ludcmp and lubksb to solve a set of equations specified in the DATA statements at the beginning of the program. The solution vector is then corrupted by the addition of random values to each component. mprove works on the corrupted vector to recover the original.

```
      PROGRAM xmprove
C     driver for routine mprove
      INTEGER N,NP
      PARAMETER(N=5,NP=5)
      INTEGER i,j,idum,indx(N)
      REAL d,a(NP,NP),b(N),x(N),aa(NP,NP),ran3
      DATA a/1.0,2.0,1.0,4.0,5.0,2.0,3.0,1.0,5.0,1.0,
```

```
      *      3.0,4.0,1.0,1.0,2.0,4.0,5.0,1.0,2.0,3.0,
      *      5.0,1.0,1.0,3.0,4.0/
             DATA b/1.0,1.0,1.0,1.0,1.0,1.0/
             do 12 i=1,N
               x(i)=b(i)
               do 11 j=1,N
                 aa(i,j)=a(i,j)
      11       continue
      12     continue
             call ludcmp(aa,N,NP,indx,d)
             call lubksb(aa,N,NP,indx,x)
             write(*,'(/1x,a)') 'Solution vector for the equations:'
             write(*,'(1x,5f12.6)') (x(i),i=1,N)
      C      now phoney up x and let mprove fix it
             idum=-13
             do 13 i=1,N
               x(i)=x(i)*(1.0+0.2*ran3(idum))
      13     continue
             write(*,'(/1x,a)') 'Solution vector with noise added:'
             write(*,'(1x,5f12.6)') (x(i),i=1,N)
             call mprove(a,aa,N,NP,indx,b,x)
             write(*,'(/1x,a)') 'Solution vector recovered by MPROVE:'
             write(*,'(1x,5f12.6)') (x(i),i=1,N)
             END
```

The pair svdcmp, svbksb is tested in the same manner as ludcmp, lubksb. That is, svdcmp is checked independently to see that it yields proper decomposition of matrices. Then the pair of programs is tested as a unit to see that it provides correct solutions to some linear sets. (Note: Because of the order of programs in *Numerical Recipes*, the test of the pair in this case comes first).

Driver xsvbksb brings in matrices a and right-hand side vectors b from MA-TRX1.DAT. Matrix a, itself, is saved for later use. It is copied into matrix u for processing by svdcmp. The results of the processing are the three arrays u, w, v which form the singular value decomposition of a. The right-hand side vectors are fed one at a time to vector c, and the resulting solution vectors x are checked for accuracy through application of the saved matrix a.

```
             PROGRAM xsvbksb
      C      driver for routine svbksb, which calls routine svdcmp
             INTEGER MP,NP
             PARAMETER(MP=20,NP=20)
             INTEGER j,k,l,m,n
             REAL a(NP,NP),b(NP,MP),u(NP,NP),w(NP)
             REAL v(NP,NP),c(NP),x(NP)
             REAL wmax,wmin
             CHARACTER dummy*3
             open(7,file='MATRX1.DAT',status='old')
      10     read(7,'(a)') dummy
             if (dummy.eq.'END') goto 99
             read(7,*)
             read(7,*) n,m
             read(7,*)
             read(7,*) ((a(k,l), l=1,n), k=1,n)
             read(7,*)
             read(7,*) ((b(k,l), k=1,n), l=1,m)
```

```
C       copy a into u
        do 12 k=1,n
          do 11 l=1,n
            u(k,l)=a(k,l)
11        continue
12      continue
C       decompose matrix a
        call svdcmp(u,n,n,NP,NP,w,v)
C       find maximum singular value
        wmax=0.0
        do 13 k=1,n
          if (w(k).gt.wmax) wmax=w(k)
13      continue
C       define "small"
        wmin=wmax*(1.0e-6)
C       zero the "small" singular values
        do 14 k=1,n
          if (w(k).lt.wmin) w(k)=0.0
14      continue
C       backsubstitute for each right-hand side vector
        do 18 l=1,m
          write(*,'(1x,a,i2)') 'Vector number ',l
          do 15 k=1,n
            c(k)=b(k,l)
15        continue
          call svbksb(u,w,v,n,n,NP,NP,c,x)
          write(*,*) '    Solution vector is:'
          write(*,'(1x,6f12.6)') (x(k), k=1,n)
          write(*,*) '    Original right-hand side vector:'
          write(*,'(1x,6f12.6)') (c(k), k=1,n)
          write(*,*) '    Result of (matrix)*(sol''n vector):'
          do 17 k=1,n
            c(k)=0.0
            do 16 j=1,n
              c(k)=c(k)+a(k,j)*x(j)
16          continue
17        continue
          write(*,'(1x,6f12.6)') (c(k), k=1,n)
18      continue
        write(*,*) '***********************************'
        write(*,*) 'Press RETURN for next problem'
        read(*,*)
        goto 10
99      close(7)
        END
```

The rather pathological test matrices for xsvdcmp are given in the Appendix as file MATRX3.DAT. The routine reads each matrix a in and passes a copy of each to svdcmp for singular value decomposition into u, w, and v. Then u, w, and the transpose of v are multiplied together. The result is compared to a saved copy of a.

```
        PROGRAM xsvdcmp
C       driver for routine svdcmp
        INTEGER MP,NP
        PARAMETER(MP=20,NP=20)
        INTEGER j,k,l,m,n
        REAL a(MP,NP),u(MP,NP),w(NP),v(NP,NP)
```

```
      CHARACTER dummy*3
      open(7,file='MATRX3.DAT',status='old')
10    read(7,'(a)') dummy
      if (dummy.eq.'END') goto 99
      read(7,*)
      read(7,*) m,n
      read(7,*)
C     copy original matrix into u
      do 12 k=1,m
        read(7,*) (a(k,l), l=1,n)
        do 11 l=1,n
          u(k,l)=a(k,l)
11      continue
12    continue
C     perform decomposition
      call svdcmp(u,m,n,MP,NP,w,v)
C     print results
      write(*,*) 'Decomposition Matrices:'
      write(*,*) 'Matrix U'
      do 13 k=1,m
        write(*,'(1x,6f12.6)') (u(k,l),l=1,n)
13    continue
      write(*,*) 'Diagonal of Matrix W'
      write(*,'(1x,6f12.6)') (w(k),k=1,n)
      write(*,*) 'Matrix V-Transpose'
      do 14 k=1,n
        write(*,'(1x,6f12.6)') (v(l,k),l=1,n)
14    continue
      write(*,*) 'Check product against original matrix:'
      write(*,*) 'Original Matrix:'
      do 15 k=1,m
        write(*,'(1x,6f12.6)') (a(k,l),l=1,n)
15    continue
      write(*,*) 'Product U*W*(V-Transpose):'
      do 18 k=1,m
        do 17 l=1,n
          a(k,l)=0.0
          do 16 j=1,n
            a(k,l)=a(k,l)+u(k,j)*w(j)*v(l,j)
16        continue
17      continue
        write(*,'(1x,6f12.6)') (a(k,l),l=1,n)
18    continue
      write(*,*) '*********************************'
      write(*,*) 'Press RETURN for next problem'
      read(*,*)
      goto 10
99    close(7)
      END
```

A *cyclic tridiagonal* matrix is just like a tridiagonal matrix, but with additional nonzero elements in the top right and bottom left corners. Sample program xcyclic generates a random cyclic tridiagonal matrix and right-hand side vector. It then calls cyclic to solve the resulting linear system. The solution is checked by also solving the system with ludcmp and lubksb and printing out the fractional discrepancy in the answer.

```
      PROGRAM xcyclic
C     driver for routine cyclic
      INTEGER N
      PARAMETER (N=20)
      REAL alpha,beta,d,ran2
      REAL a(N),b(N),c(N),r(N),x(N),aa(N,N)
      INTEGER indx(N)
      INTEGER i,j,idum
      idum= -23
      do 12 i=1,N
        do 11 j=1,N
          aa(i,j)=0.
11      continue
12    continue
      do 13 i=1,N
        b(i)=ran2(idum)
        aa(i,i)=b(i)
        r(i)=ran2(idum)
13    continue
      do 14 i=1,N-1
        a(i+1)=ran2(idum)
        aa(i+1,i)=a(i+1)
        c(i)=ran2(idum)
        aa(i,i+1)=c(i)
14    continue
      alpha=ran2(idum)
      aa(N,1)=alpha
      beta=ran2(idum)
      aa(1,N)=beta
      call cyclic(a,b,c,alpha,beta,r,x,N)
      call ludcmp(aa,N,N,indx,d)
      call lubksb(aa,N,N,indx,r)
      do 15 i=1,N
        write(*,*) i,(x(i)-r(i))/(x(i)+r(i))
15    continue
      END
```

Next follow the routines for testing the sparse matrix routines. The first, xsprsin, takes the matrix defined in Equation (2.7.27) of *Numerical Recipes*, prints out the corresponding ija and sa vectors (which you can check against Equation 2.7.28), and then reconstructs the matrix from these vectors.

```
      PROGRAM xsprsin
C     driver for routine sprsin
      INTEGER NP,NMAX
      PARAMETER(NP=5,NMAX=2*NP*NP+1)
      INTEGER i,j,msize,ija(NMAX)
      REAL a(NP,NP),aa(NP,NP),sa(NMAX)
      DATA a/3.0,0.0,0.0,0.0,0.0,
     *      0.0,4.0,7.0,0.0,0.0,
     *      1.0,0.0,5.0,0.0,0.0,
     *      0.0,0.0,9.0,0.0,6.0,
     *      0.0,0.0,0.0,2.0,5.0/
      call sprsin(a,NP,NP,0.5,NMAX,sa,ija)
      msize=ija(ija(1)-1)-1
      sa(NP+1)=0.0
      write(*,'(t4,a,t18,a,t24,a)') 'index','ija','sa'
```

```
      do 11 i=1,msize
        write(*,'(t2,i4,t16,i4,t20,f12.6)') i,ija(i),sa(i)
11    continue
      do 13 i=1,NP
        do 12 j=1,NP
          aa(i,j)=0.0
12      continue
13    continue
      do 15 i=1,NP
        aa(i,i)=sa(i)
        do 14 j=ija(i),ija(i+1)-1
          aa(i,ija(j))=sa(j)
14      continue
15    continue
      write(*,*) 'Original matrix:'
      write(*,'(5f7.1)') ((a(i,j),j=1,NP),i=1,NP)
      write(*,*) 'Reconstructed matrix:'
      write(*,'(5f7.1)') ((aa(i,j),j=1,NP),i=1,NP)
      END
```

Program xsprsax takes the same sparse matrix and multiplies it by the vector (1,2,3,4,5) using sprsax. It compares the result to that of a full matrix multiplication.

```
      PROGRAM xsprsax
C     driver for routine sprsax
      INTEGER NP,NMAX
      PARAMETER(NP=5,NMAX=2*NP*NP+1)
      INTEGER i,j,msize,ija(NMAX)
      REAL a(NP,NP),sa(NMAX),ax(NP),b(NP),x(NP)
      DATA a/3.0,0.0,0.0,0.0,0.0,
     *       0.0,4.0,7.0,0.0,0.0,
     *       1.0,0.0,5.0,0.0,0.0,
     *       0.0,0.0,9.0,0.0,6.0,
     *       0.0,0.0,0.0,2.0,5.0/
      DATA x/1.0,2.0,3.0,4.0,5.0/
      call sprsin(a,NP,NP,0.5,NMAX,sa,ija)
      msize=ija(1)-2
      call sprsax(sa,ija,x,b,msize)
      do 12 i=1,msize
        ax(i)=0.0
        do 11 j=1,msize
          ax(i)=ax(i)+a(i,j)*x(j)
11      continue
12    continue
      write(*,'(t4,a,t18,a)') 'Reference','sprsax result'
      do 13 i=1,msize
        write(*,'(t4,f5.2,t22,f5.2)') ax(i),b(i)
13    continue
      END
```

Program xsprstx works in exactly the same way as the previous program to test sprstx by computing the product of the transpose of the sparse matrix with the vector:

```
      PROGRAM xsprstx
C     driver for routine sprstx
      INTEGER NP,NMAX
```

```
      PARAMETER(NP=5,NMAX=2*NP*NP+1)
      INTEGER i,j,msize,ija(NMAX)
      REAL a(NP,NP),sa(NMAX),ax(NP),b(NP),x(NP)
      DATA a/3.0,0.0,0.0,0.0,0.0,
     *    0.0,4.0,7.0,0.0,0.0,
     *    1.0,0.0,5.0,0.0,0.0,
     *    0.0,0.0,9.0,0.0,6.0,
     *    0.0,0.0,0.0,2.0,5.0/
      DATA x/1.0,2.0,3.0,4.0,5.0/
      call sprsin(a,NP,NP,0.5,NMAX,sa,ija)
      msize=ija(1)-2
      call sprstx(sa,ija,x,b,msize)
      do 12 i=1,msize
        ax(i)=0.0
        do 11 j=1,msize
          ax(i)=ax(i)+a(j,i)*x(j)
11      continue
12    continue
      write(*,'(t4,a,t18,a)') 'Reference','sprstx result'
      do 13 i=1,msize
        write(*,'(t4,f5.2,t22,f5.2)') ax(i),b(i)
13    continue
      END
```

Program xsprstp simply prints out the transpose of our sample sparse matrix after calling sprstp:

```
      PROGRAM xsprstp
C     driver for routine sprstp
      INTEGER NP,NMAX
      PARAMETER(NP=5,NMAX=2*NP*NP+1)
      INTEGER i,j,ija(NMAX),ijat(NMAX)
      REAL a(NP,NP),at(NP,NP),sa(NMAX),sat(NMAX)
      DATA a/3.0,0.0,0.0,0.0,0.0,
     *    0.0,4.0,7.0,0.0,0.0,
     *    1.0,0.0,5.0,0.0,0.0,
     *    0.0,0.0,9.0,0.0,6.0,
     *    0.0,0.0,0.0,2.0,5.0/
      call sprsin(a,NP,NP,0.5,NMAX,sa,ija)
      call sprstp(sa,ija,sat,ijat)
      do 12 i=1,NP
        do 11 j=1,NP
          at(i,j)=0.0
11      continue
12    continue
      do 14 i=1,NP
        at(i,i)=sat(i)
        do 13 j=ijat(i),ijat(i+1)-1
          at(i,ijat(j))=sat(j)
13      continue
14    continue
      write(*,*) 'Original matrix:'
      write(*,'(5f7.1)') ((a(i,j),j=1,NP),i=1,NP)
      write(*,*) 'Transpose:'
      write(*,'(5f7.1)') ((at(i,j),j=1,NP),i=1,NP)
      END
```

The next program, xsprspm, demonstrates the use of sprspm. Two sparse matrices **A** and **B** are defined, and then the matrix **A** · **B** is computed. Since sprspm actually calculates **A** · **B**^{T}, we first form **B**^{T} by a call to sprstp. Only a given sparsity pattern is output by sprspm. In the example, we specify this to be the tridiagonal part of **A** · **B**. As usual, we check the answer by carrying out a full matrix multiplication.

```
      PROGRAM xsprspm
C     driver for routine sprspm
      INTEGER NP,NMAX
      PARAMETER(NP=5,NMAX=2*NP*NP+1)
      INTEGER i,j,k,ija(NMAX),ijb(NMAX),ijbt(NMAX),ijc(NMAX)
      REAL sa(NMAX),sb(NMAX),sbt(NMAX),sc(NMAX)
      REAL a(NP,NP),b(NP,NP),c(NP,NP),ab(NP,NP)
      DATA a/1.0,0.5,0.0,0.0,0.0,
     *      0.5,2.0,0.5,0.0,0.0,
     *      0.0,0.5,3.0,0.5,0.0,
     *      0.0,0.0,0.5,4.0,0.5,
     *      0.0,0.0,0.0,0.5,5.0/
      DATA b/1.0,1.0,0.0,0.0,0.0,
     *      1.0,2.0,1.0,0.0,0.0,
     *      0.0,1.0,3.0,1.0,0.0,
     *      0.0,0.0,1.0,4.0,1.0,
     *      0.0,0.0,0.0,1.0,5.0/
      call sprsin(a,NP,NP,0.5,NMAX,sa,ija)
      call sprsin(b,NP,NP,0.5,NMAX,sb,ijb)
      call sprstp(sb,ijb,sbt,ijbt)
C     specify tridiagonal output, using fact that a is tridiagonal
      do 11 i=1,ija(ija(1)-1)-1
        ijc(i)=ija(i)
11    continue
      call sprspm(sa,ija,sbt,ijbt,sc,ijc)
      do 14 i=1,NP
        do 13 j=1,NP
          ab(i,j)=0.0
          do 12 k=1,NP
            ab(i,j)=ab(i,j)+a(i,k)*b(k,j)
12        continue
13      continue
14    continue
      write(*,*) 'Reference matrix:'
      write(*,'(5f7.1)') ((ab(i,j),j=1,NP),i=1,NP)
      write(*,*) 'sprspm matrix (should show only tridiagonals):'
      do 16 i=1,NP
        do 15 j=1,NP
          c(i,j)=0.0
15      continue
16    continue
      do 18 i=1,NP
        c(i,i)=sc(i)
        do 17 j=ijc(i),ijc(i+1)-1
          c(i,ijc(j))=sc(j)
17      continue
18    continue
      write(*,'(5f7.1)') ((c(i,j),j=1,NP),i=1,NP)
      END
```

Routine sprstm operates in the same way as sprspm, except the output is all components of $\mathbf{A} \cdot \mathbf{B}^T$ greater than some threshold value, here chosen to be 0.99. The test program xsprstm is thus very similar to the previous test program.

```
      PROGRAM xsprstm
C     driver for routine sprstm
      INTEGER NP,NMAX
      REAL THRESH
      PARAMETER(NP=5,NMAX=2*NP*NP+1,THRESH=0.99)
      INTEGER i,j,k,ija(NMAX),ijb(NMAX),ijbt(NMAX),ijc(NMAX),msize
      REAL sa(NMAX),sb(NMAX),sbt(NMAX),sc(NMAX)
      REAL a(NP,NP),b(NP,NP),c(NP,NP),ab(NP,NP)
      DATA a/1.0,0.5,0.0,0.0,0.0,
     *    0.5,2.0,0.5,0.0,0.0,
     *    0.0,0.5,3.0,0.5,0.0,
     *    0.0,0.0,0.5,4.0,0.5,
     *    0.0,0.0,0.0,0.5,5.0/
      DATA b/1.0,1.0,0.0,0.0,0.0,
     *    1.0,2.0,1.0,0.0,0.0,
     *    0.0,1.0,3.0,1.0,0.0,
     *    0.0,0.0,1.0,4.0,1.0,
     *    0.0,0.0,0.0,1.0,5.0/
      call sprsin(a,NP,NP,0.5,NMAX,sa,ija)
      call sprsin(b,NP,NP,0.5,NMAX,sb,ijb)
      call sprstp(sb,ijb,sbt,ijbt)
      msize=ija(ija(1)-1)-1
      call sprstm(sa,ija,sbt,ijbt,THRESH,msize,sc,ijc)
      do 13 i=1,NP
        do 12 j=1,NP
          ab(i,j)=0.0
          do 11 k=1,NP
            ab(i,j)=ab(i,j)+a(i,k)*b(k,j)
11        continue
12      continue
13    continue
      write(*,*) 'Reference matrix:'
      write(*,'(5f7.1)') ((ab(i,j),j=1,NP),i=1,NP)
      write(*,*) 'sprstm matrix (off-diag. elements of mag >):',THRESH
      do 15 i=1,NP
        do 14 j=1,NP
          c(i,j)=0.0
14      continue
15    continue
      do 17 i=1,NP
        c(i,i)=sc(i)
        do 16 j=ijc(i),ijc(i+1)-1
          c(i,ijc(j))=sc(j)
16      continue
17    continue
      write(*,'(5f7.1)') ((c(i,j),j=1,NP),i=1,NP)
      END
```

Routine linbcg solves linear systems $\mathbf{A} \cdot \mathbf{x} = \mathbf{b}$ with a sparse matrix $\mathbf{A}$. The matrix $\mathbf{A}$ must be specified in sparse matrix format as described in *Numerical Recipes*. In our sample program xlinbcg we define $\mathbf{A}$ to be the 20×20 matrix

$$\begin{pmatrix} 1.0 & 2.0 & 0.0 & 0.0 & \cdots \\ -2.0 & 1.0 & 2.0 & 0.0 & \cdots \\ 0.0 & -2.0 & 1.0 & 2.0 & \cdots \\ 0.0 & 0.0 & -2.0 & 1.0 & \cdots \\ \vdots & \vdots & \vdots & \vdots & \ddots \end{pmatrix}$$

As a right-hand side vector $\mathbf{b}$ we have taken $(3.0, 1.0, 1.0, \ldots, -1.0)$, and the solution is given as $\mathbf{x}$. Notice that the components of $\mathbf{x}$ are all initialized to zero. You will set them to some initial guess of the solution to your own problem, but a zero guess will usually suffice. The solution in `xlinbcg` is given the usual checks.

```
      PROGRAM xlinbcg
C     driver for routine linbcg
      INTEGER NP,NMAX,ITOL,ITMAX
      DOUBLE PRECISION TOL
      PARAMETER (NP=20,NMAX=1000,ITOL=1,TOL=1.d-9,ITMAX=75)
      INTEGER i,iter,ija
      DOUBLE PRECISION b(NP),x(NP),bcmp(NP),sa,err
      COMMON /mat/ sa(NMAX),ija(NMAX)
      do 11 i=1,NP
        x(i)=0.d0
        b(i)=1.d0
11    continue
      b(1)=3.d0
      b(NP)=-1.d0
      call linbcg(NP,b,x,ITOL,TOL,ITMAX,iter,err)
      write(*,'(/1x,a,e15.6)') 'Estimated error:',err
      write(*,'(/1x,a,i6)') 'Iterations needed:',iter
      write(*,'(/1x,a)') 'Solution vector:'
      write(*,'(1x,5f12.6)') (x(i),i=1,NP)
      call dsprsax(sa,ija,x,bcmp,NP)
C     this is a double precision version of sprsax
      write(*,'(/1x,a)') 'press RETURN to continue...'
      read(*,*)
      write(*,'(1x,a/t8,a,t22,a)') 'Test of solution vector:','a*x','b'
      do 12 i=1,NP
        write(*,'(1x,2f12.6)') bcmp(i),b(i)
12    continue
      END

      BLOCK DATA
      INTEGER NMAX
      PARAMETER(NMAX=1000)
      DOUBLE PRECISION sa
      INTEGER ija
      COMMON /mat/ sa(NMAX),ija(NMAX)
C     logical length = 59
      DATA sa/3.d0,3.d0,3.d0,3.d0,3.d0,3.d0,3.d0,3.d0,
     *    3.d0,3.d0,3.d0,3.d0,3.d0,3.d0,3.d0,3.d0,
     *    3.d0,3.d0,3.d0,3.d0,0.d0,2.d0,-2.d0,2.d0,
     *    -2.d0,2.d0,-2.d0,2.d0,-2.d0,2.d0,-2.d0,
     *    2.d0,-2.d0,2.d0,-2.d0,2.d0,-2.d0,2.d0,-2.d0,
     *    2.d0,-2.d0,2.d0,-2.d0,2.d0,-2.d0,2.d0,
     *    -2.d0,2.d0,-2.d0,2.d0,-2.d0,2.d0,-2.d0,
     *    2.d0,-2.d0,2.d0,-2.d0,2.d0,-2.d0,941*0.d0/
      DATA ija/22,23,25,27,29,31,33,35,37,39,41,43,45,47,49,51,53,
```

```
      *    55,57,59,60,2,1,3,2,4,3,5,4,6,5,7,6,8,7,9,8,10,9,11,10,
      *    12,11,13,12,14,13,15,14,16,15,17,16,18,17,19,18,20,19,941*0/
      END
```

Vandermonde matrices of dimension $N \times N$ have elements that are entirely integer powers of N arbitrary numbers $x_1 \ldots x_N$. (See *Numerical Recipes* for details). In the demonstration program xvander we provide five such numbers to specify a 5×5 matrix, and five elements of a right-hand side vector Q. Routine vander is used to find the solution vector W. This vector is tested by applying the matrix to W and comparing the result to Q.

```
      PROGRAM xvander
C     driver for routine vander
      INTEGER N
      PARAMETER(N=5)
      INTEGER i,j
      DOUBLE PRECISION sum,x(N),q(N),w(N),term(N)
      DATA x/1.0d0,1.5d0,2.0d0,2.5d0,3.0d0/
      DATA q/1.0d0,1.5d0,2.0d0,2.5d0,3.0d0/
      call vander(x,w,q,N)
      write(*,*) 'Solution vector:'
      do 11 i=1,N
        write(*,'(5x,a2,i1,a4,e12.6)') 'W(',i,') = ',w(i)
11    continue
      write(*,'(/1x,a)') 'Test of solution vector:'
      write(*,'(1x,t6,a,t19,a)') 'mtrx*sol''n','original'
      sum=0.0
      do 12 i=1,N
        term(i)=w(i)
        sum=sum+w(i)
12    continue
      write(*,'(1x,2f12.4)') sum,q(1)
      do 14 i=2,N
        sum=0.0
        do 13 j=1,N
          term(j)=term(j)*x(j)
          sum=sum+term(j)
13      continue
        write(*,'(1x,2f12.4)') sum,q(i)
14    continue
      END
```

A very similar test is applied to toeplz, which operates on Toeplitz matrices. The $N \times N$ Toeplitz matrix is specified by $2N - 1$ numbers r_i, in this case taken to be simply a linear progression of values. A right-hand side y_i is chosen likewise. toeplz finds the solution vector x_i, and checks it in the usual fashion.

```
      PROGRAM xtoeplz
C     driver for routine toeplz
      INTEGER N,N2
      PARAMETER(N=5,N2=2*N)
      INTEGER i,j
      REAL sum,x(N),y(N),r(N2)
      do 11 i=1,N
        y(i)=0.1*i
11    continue
```

```
      do 12 i=1,2*N-1
        r(i)=1./i
12    continue
      call toeplz(r,x,y,N)
      write(*,*) 'Solution vector:'
      do 13 i=1,N
        write(*,'(5x,a2,i1,a4,e13.6)') 'X(',i,') = ',x(i)
13    continue
      write(*,'(/1x,a)') 'Test of solution:'
      write(*,'(1x,t6,a,t19,a)') 'mtrx*soln','original'
      do 15 i=1,N
        sum=0.0
        do 14 j=1,N
          sum=sum+r(N+i-j)*x(j)
14      continue
        write(*,'(1x,2f12.4)') sum,y(i)
15    continue
      END
```

The routines `choldc` and `cholsl` are tested together by `xcholsl`. First we form a 3×3 positive-definite, symmetric matrix. We compute its Cholesky decomposition, and check the answer by explicitly multiplying back the Cholesky factors. We then use the decomposition to solve a 3×3 system of equations, and check the answer by substituting back in the original system.

```
      PROGRAM xcholsl
C     driver for routine cholsl
      INTEGER N
      PARAMETER (N=3)
      INTEGER i,j,k
      REAL sum,a(N,N),aorig(N,N),atest(N,N),chol(N,N),p(N),b(N),x(N)
      DATA a/100.,15.,.01,15.,2.3,.01,.01,.01,1./
      DATA b/.4,.02,99./
      do 12 i=1,N
        do 11 j=1,N
          aorig(i,j)=a(i,j)
11      continue
12    continue
      call choldc(a,N,N,p)
      do 14 i=1,N
        do 13 j=1,N
          if (i.gt.j) then
            chol(i,j)=a(i,j)
          else if (i.eq.j) then
            chol(i,j)=p(i)
          else
            chol(i,j)=0.
          endif
13      continue
14    continue
      do 17 i=1,N
        do 16 j=1,N
          sum=0.
          do 15 k=1,N
            sum=sum+chol(i,k)*chol(j,k)
15        continue
          atest(i,j)=sum
```

```
16        continue
17        continue
          write(*,*) 'Original matrix:'
          write(*,100) ((aorig(i,j),j=1,N),i=1,N)
          write(*,*)
          write(*,*) 'Product of Cholesky factors:'
          write(*,100) ((atest(i,j),j=1,N),i=1,N)
100       format(1p3e16.6)
          write(*,*)
          call cholsl(a,N,N,p,b,x)
          do 19 i=1,N
            sum=0.
            do 18 j=1,N
              sum=sum+aorig(i,j)*x(j)
18          continue
            p(i)=sum
19        continue
          write(*,*) 'Check solution vector:'
          write(*,101) (p(i),b(i),i=1,N)
101       format(1p2e16.6)
          END
```

The QR decomposition routine qrdcmp is tested by a program xqrdcmp that is very similar to the program for testing the LU decomposition. It reads in the same set of test matrices from the file MATRX1.DAT, does the decomposition of each one, and then checks the result by multiplying the **Q** and **R** matrices.

```
      PROGRAM xqrdcmp
C     driver for routine qrdcmp
      INTEGER NP
      PARAMETER(NP=20)
      INTEGER i,j,k,l,m,n
      REAL con,a(NP,NP),c(NP),d(NP),q(NP,NP),qt(NP,NP),r(NP,NP),
     *     x(NP,NP)
      CHARACTER txt*3
      LOGICAL sing
      open(7,file='MATRX1.DAT',status='old')
      read(7,*)
10    read(7,*)
      read(7,*) n,m
      read(7,*)
      read(7,*) ((a(k,l), l=1,n), k=1,n)
      read(7,*)
      read(7,*) ((x(k,l), k=1,n), l=1,m)
C     print out a-matrix for comparison with product of Q and R
C     decomposition matrices.
      write(*,*) 'Original matrix:'
      do 11 k=1,n
        write(*,'(1x,6f12.6)') (a(k,l), l=1,n)
11    continue
C     perform the decomposition
      call qrdcmp(a,n,NP,c,d,sing)
      if (sing) write(*,*) 'Singularity in QR decomposition.'
C     find the Q and R matrices
      do 13 k=1,n
        do 12 l=1,n
          if (l.gt.k) then
```

```
          r(k,l)=a(k,l)
          q(k,l)=0.0
        else if (l.lt.k) then
          r(k,l)=0.0
          q(k,l)=0.0
        else
          r(k,l)=d(k)
          q(k,l)=1.0
        endif
12      continue
13    continue
      do 21 i=n-1,1,-1
        con=0.0
        do 14 k=i,n
          con=con+a(k,i)**2
14      continue
        con=con/2.0
        do 17 k=i,n
          do 16 l=i,n
            qt(k,l)=0.0
            do 15 j=i,n
              qt(k,l)=qt(k,l)+q(j,l)*a(k,i)*a(j,i)/con
15          continue
16        continue
17      continue
        do 19 k=i,n
          do 18 l=i,n
            q(k,l)=q(k,l)-qt(k,l)
18        continue
19      continue
21    continue
C     compute product of Q and R matrices for comparison with original matrix.
      do 24 k=1,n
        do 23 l=1,n
          x(k,l)=0.0
          do 22 j=1,n
            x(k,l)=x(k,l)+q(k,j)*r(j,l)
22        continue
23      continue
24    continue
      write(*,*) 'Product of Q and R matrices:'
      do 25 k=1,n
        write(*,'(1x,6f12.6)') (x(k,l), l=1,n)
25    continue
      write(*,*) 'Q matrix of the decomposition:'
      do 26 k=1,n
        write(*,'(1x,6f12.6)') (q(k,l), l=1,n)
26    continue
      write(*,*) 'R matrix of the decomposition:'
      do 27 k=1,n
        write(*,'(1x,6f12.6)') (r(k,l), l=1,n)
27    continue
      write(*,*) '*********************************'
      write(*,*) 'Press RETURN for next problem:'
      read(*,*)
      read(7,'(a3)') txt
      if (txt.ne.'END') goto 10
```

```
      close(7)
      END
```

Similarly routine `qrsolv` is tested by a program `xqrsolv` that is almost the same as the test program for LU backsubstitution, and uses the same set of test matrices in the file `MATRX1.DAT`.

```
      PROGRAM xqrsolv
C     driver for routine qrsolv
      INTEGER NP
      PARAMETER (NP=20)
      REAL a(NP,NP),ai(NP,NP),b(NP,NP),c(NP),d(NP),x(NP)
      INTEGER j,k,l,m,n
      CHARACTER txt*3
      LOGICAL sing
      open(7,file='MATRX1.DAT',status='old')
      read(7,*)
10    read(7,*)
      read(7,*) n,m
      read(7,*)
      read(7,*) ((a(k,l), l=1,n), k=1,n)
      read(7,*)
      read(7,*) ((b(k,l), k=1,n), l=1,m)
C     save matrix a for later testing
      do 12 l=1,n
        do 11 k=1,n
          ai(k,l)=a(k,l)
11      continue
12    continue
C     do qr decomposition
      call qrdcmp(a,n,NP,c,d,sing)
      if (sing) write(*,*) 'Singularity in QR decomposition.'
C     solve equations for each right-hand vector
      do 16 k=1,m
        do 13 l=1,n
          x(l)=b(l,k)
13      continue
        call qrsolv(a,n,NP,c,d,x)
C     test results with original matrix
        write(*,*) 'Right-hand side vector:'
        write(*,'(1x,6f12.6)') (b(l,k), l=1,n)
        write(*,*) 'Result of matrix applied to sol''n vector'
        do 15 l=1,n
          b(l,k)=0.0
          do 14 j=1,n
            b(l,k)=b(l,k)+ai(l,j)*x(j)
14        continue
15      continue
        write(*,'(1x,6f12.6)') (b(l,k), l=1,n)
        write(*,*) '*********************************'
16    continue
      write(*,*) 'Press RETURN for next problem:'
      read(*,*)
      read(7,'(a3)') txt
      if (txt.ne.'END') goto 10
      close(7)
      END
```

The test program for `qrupdt` once again reads in test matrices from the file
`MATRX1.DAT`. However, it uses the solution vectors in the file as vectors defining the
update to be performed. The solution is checked by explicitly updating the original
matrices and then doing an *ab initio* QR decomposition.

```
      PROGRAM xqrupdt
C     driver for routine qrdupd
      INTEGER NP
      PARAMETER(NP=20)
      INTEGER i,j,k,l,m,n
      REAL con,a(NP,NP),au(NP,NP),c(NP),d(NP),q(NP,NP),qt(NP,NP),
     *    r(NP,NP),s(NP,NP),u(NP),v(NP),x(NP,NP)
      CHARACTER txt*3
      LOGICAL sing
      open(7,file='MATRX1.DAT',status='old')
      read(7,*)
10    read(7,*)
      read(7,*) n,m
      read(7,*)
      read(7,*) ((a(k,l), l=1,n), k=1,n)
      read(7,*)
      read(7,*) ((s(k,l), k=1,n), l=1,m)
C     print out a-matrix for comparison with product of Q and R
C     decomposition matrices.
      write(*,*) 'Original matrix:'
      do 11 k=1,n
        write(*,'(1x,6f12.6)') (a(k,l), l=1,n)
11    continue
C     updated matrix we'll use later
      do 13 k=1,n
        do 12 l=1,n
          au(k,l)=a(k,l)+s(k,1)*s(1,2)
12      continue
13    continue
C     perform the initial decomposition
      call qrdcmp(a,n,NP,c,d,sing)
      if (sing) write(*,*) 'Singularity in QR decomposition.'
C     find the Q and R matrices
      do 15 k=1,n
        do 14 l=1,n
          if (l.gt.k) then
            r(k,l)=a(k,l)
            q(k,l)=0.0
          else if (l.lt.k) then
            r(k,l)=0.0
            q(k,l)=0.0
          else
            r(k,l)=d(k)
            q(k,l)=1.0
          endif
14      continue
15    continue
      do 23 i=n-1,1,-1
        con=0.0
        do 16 k=i,n
          con=con+a(k,i)**2
16      continue
```

```
          con=con/2.0
          do 19 k=i,n
            do 18 l=i,n
              qt(k,l)=0.0
              do 17 j=i,n
                qt(k,l)=qt(k,l)+q(j,l)*a(k,i)*a(j,i)/con
17            continue
18          continue
19        continue
          do 22 k=i,n
            do 21 l=i,n
              q(k,l)=q(k,l)-qt(k,l)
21          continue
22        continue
23      continue
C       compute product of Q and R matrices for comparison with original matrix.
        do 26 k=1,n
          do 25 l=1,n
            x(k,l)=0.0
            do 24 j=1,n
              x(k,l)=x(k,l)+q(k,j)*r(j,l)
24          continue
25        continue
26      continue
        write(*,*) 'Product of Q and R matrices:'
        do 27 k=1,n
          write(*,'(1x,6f12.6)') (x(k,l), l=1,n)
27      continue
        write(*,*) 'Q matrix of the decomposition:'
        do 28 k=1,n
          write(*,'(1x,6f12.6)') (q(k,l), l=1,n)
28      continue
        write(*,*) 'R matrix of the decomposition:'
        do 29 k=1,n
          write(*,'(1x,6f12.6)') (r(k,l), l=1,n)
29      continue
C       Q transpose
        do 32 k=1,n
          do 31 l=1,n
            qt(k,l)=q(l,k)
31        continue
32      continue
        do 34 k=1,n
          v(k)=s(k,2)
          u(k)=0.0
          do 33 l=1,n
            u(k)=u(k)+qt(k,l)*s(l,1)
33        continue
34      continue
        call qrupdt(r,qt,n,NP,u,v)
        do 37 k=1,n
          do 36 l=1,n
            x(k,l)=0.0
            do 35 j=1,n
              x(k,l)=x(k,l)+qt(j,k)*r(j,l)
35          continue
36        continue
```

```
37       continue
         write(*,*) 'Updated matrix:'
         do 38 k=1,n
         write(*,'(1x,6f12.6)') (au(k,l), l=1,n)
38       continue
         write(*,*) 'Product of new Q and R matrices:'
         do 39 k=1,n
           write(*,'(1x,6f12.6)') (x(k,l), l=1,n)
39         continue
         write(*,*) 'New Q matrix'
         do 41 k=1,n
         write(*,'(1x,6f12.6)') (qt(l,k), l=1,n)
41       continue
         write(*,*) 'New R matrix'
         do 42 k=1,n
         write(*,'(1x,6f12.6)') (r(k,l), l=1,n)
42       continue
         write(*,*) '**********************************'
         write(*,*) 'Press RETURN for next problem:'
         read(*,*)
         read(7,'(a3)') txt
         if (txt.ne.'END') goto 10
         close(7)
         END
```

Appendix

File MATRX1.DAT:

```
MATRICES FOR INPUT TO TEST ROUTINES
Size of matrix (NxN), Number of solutions:
3,2
Matrix A:
1.0 0.0 0.0
0.0 2.0 0.0
0.0 0.0 3.0
Solution vectors:
1.0 0.0 0.0
1.0 1.0 1.0
NEXT PROBLEM
Size of matrix (NxN), Number of solutions:
3,2
Matrix A:
1.0 2.0 3.0
2.0 2.0 3.0
3.0 3.0 3.0
Solution vectors:
1.0 1.0 1.0
1.0 2.0 3.0
NEXT PROBLEM:
Size of matrix (NxN), Number of solutions:
5,2
Matrix A:
1.0 2.0 3.0 4.0 5.0
2.0 3.0 4.0 5.0 1.0
3.0 4.0 5.0 1.0 2.0
4.0 5.0 1.0 2.0 3.0
```

```
5.0 1.0 2.0 3.0 4.0
Solution vectors:
1.0 1.0 1.0 1.0 1.0
1.0 2.0 3.0 4.0 5.0
NEXT PROBLEM:
Size of matrix (NxN), Number of solutions:
5,2
Matrix A:
1.4 2.1 2.1 7.4 9.6
1.6 1.5 1.1 0.7 5.0
3.8 8.0 9.6 5.4 8.8
4.6 8.2 8.4 0.4 8.0
2.6 2.9 0.1 9.6 7.7
Solution vectors:
1.1 1.6 4.7 9.1 0.1
4.0 9.3 8.4 0.4 4.1
END
```

File MATRX2.DAT:

```
TRIDIAGONAL MATRICES FOR PROGRAM 'TRIDAG'
Dimension of matrix
3
Diagonal elements (N)
1.0 2.0 3.0
Super-diagonal elements (N-1)
2.0 3.0
Sub-diagonal elements (N-1)
2.0 3.0
Right-hand side vector (N)
1.0 2.0 3.0
NEXT PROBLEM:
Dimension of matrix
5
Diagonal elements (N)
1.0 1.0 1.0 1.0 1.0
Super-diagonal elements (N-1)
1.0 2.0 3.0 4.0
Sub-diagonal elements (N-1)
2.0 3.0 4.0 5.0
Right-hand side vector (N)
1.0 2.0 3.0 4.0 5.0
NEXT PROBLEM:
Dimension of matrix
5
Diagonal elements (N)
1.0 2.0 3.0 4.0 5.0
Super-diagonal elements (N-1)
2.0 3.0 4.0 5.0
Sub-diagonal elements (N-1)
2.0 3.0 4.0 5.0
Right-hand side vector (N)
1.0 1.0 1.0 1.0 1.0
NEXT PROBLEM:
Dimension of matrix
6
Diagonal elements (N)
```

```
9.7 9.5 5.2 3.5 5.1 6.0
Super-diagonal elements (N-1)
6.0 1.2 0.7 3.0 1.5
Sub-diagonal elements (N-1)
2.1 9.4 3.3 7.5 8.8
Right-hand side vector (N)
2.0 7.5 0.6 7.4 9.8 8.8
END
```

File MATRX3.DAT:

```
TEST MATRICES FOR SVDCMP:
Number of Rows, Columns
5,3
Matrix
1.0 2.0 3.0
2.0 3.0 4.0
3.0 4.0 5.0
4.0 5.0 6.0
5.0 6.0 7.0
NEXT PROBLEM:
Number of Rows, Columns
5,5
Matrix
1.0 2.0 3.0 4.0 5.0
2.0 2.0 3.0 4.0 5.0
3.0 3.0 3.0 4.0 5.0
4.0 4.0 4.0 4.0 5.0
5.0 5.0 5.0 5.0 5.0
NEXT PROBLEM:
Number of Rows, Columns
6,6
Matrix
3.0 5.3 5.6 3.5 6.8 5.7
0.4 8.2 6.7 1.9 2.2 5.3
7.8 8.3 7.7 3.3 1.9 4.8
5.5 8.8 3.0 1.0 5.1 6.4
5.1 5.1 3.6 5.8 5.7 4.9
3.5 2.7 5.7 8.2 9.6 2.9
END
```

Chapter 3: Interpolation and Extrapolation

Chapter 3 of Numerical Recipes deals with interpolation and extrapolation (the same routines are usable for both). Three fundamental interpolation methods are first discussed,

1. Polynomial interpolation (polint),

2. Rational function interpolation (ratint), and

3. Cubic spline interpolation (spline, splint).

To find the place in an ordered table at which to perform an interpolation, two routines are given, locate and hunt. Also, for cases in which the actual coefficients of a polynomial interpolation are desired, the routines polcoe and polcof are provided (along with important warnings circumscribing their usefulness).

For higher-dimensional interpolations, Numerical Recipes treats only problems on a regularly spaced grid. Routine polin2 does a two-dimensional polynomial interpolation that aims at accuracy rather than smoothness. When smooth interpolation is desired, the methods shown in bcucof and bcuint for bicubic interpolation are recommended. In the case of two-dimensional spline interpolations, the routines splie2 and splin2 are offered.

$$\star \quad \star \quad \star \quad \star$$

Program polint takes two arrays xa and ya of length N that express the known values of a function, and calculates the value, at a point x, of the unique polynomial of degree $N-1$ passing through all the given values. For the purpose of illustration, in xpolint we have taken evenly spaced xa(i) and set ya(i) equal to simple functions (sines and exponentials) of these xa(i). For the sine we use an interval of length π, and for the exponential an interval of length 1.0. You may choose the number N of reference points and observe the improvement of the results as N increases. The test points x are slightly shifted from the reference points so that you can compare the estimated error dy with the actual error. By removing the shift, you may check that the polynomial actually hits all reference points.

```
      PROGRAM xpolint
C     driver for routine polint
      INTEGER NP
      REAL PI
      PARAMETER(NP=10,PI=3.1415926)
      INTEGER i,n,nfunc
      REAL dy,f,x,y,xa(NP),ya(NP)
      write(*,*) 'Generation of interpolation tables'
```

```
      write(*,*) ' ... sin(x)    0<x<PI'
      write(*,*) ' ... exp(x)    0<x<1 '
      write(*,*) 'How many entries go in these tables? (note: N<10)'
      read(*,*) n
      do 14 nfunc=1,2
        if (nfunc.eq.1) then
          write(*,*) 'sine function from 0 to PI'
          do 11 i=1,n
            xa(i)=i*PI/n
            ya(i)=sin(xa(i))
11        continue
        else if (nfunc.eq.2) then
          write(*,*) 'exponential function from 0 to 1'
          do 12 i=1,n
            xa(i)=i*1.0/n
            ya(i)=exp(xa(i))
12        continue
        else
          stop
        endif
        write(*,'(t10,a1,t20,a4,t28,a12,t46,a5)')
     *       'x','f(x)','interpolated','error'
        do 13 i=1,10
          if (nfunc.eq.1) then
            x=(-0.05+i/10.0)*PI
            f=sin(x)
          else if (nfunc.eq.2) then
            x=(-0.05+i/10.0)
            f=exp(x)
          endif
          call polint(xa,ya,n,x,y,dy)
          write(*,'(1x,3f12.6,e15.4)') x,f,y,dy
13      continue
        write(*,*) '*********************************'
        write(*,*) 'Press RETURN'
        read(*,*)
14    continue
      END
```

`ratint` is functionally similar to `polint` in that it also returns a value y for the function at point x, and an error estimate dy as well. In this case the values are determined from the unique diagonal rational function that passes through all the reference points. If you inspect the driver closely, you will find that two of the test points fall directly on top of reference points and should give exact results. The remainder do not. You can compare the estimated error dyy to the actual error $|yy - yexp|$ for these cases.

```
      PROGRAM xratint
C     driver for routine ratint
      INTEGER NPT
      REAL EPSSQ
      PARAMETER(NPT=6,EPSSQ=1.0)
      INTEGER i
      REAL dyy,f,xx,yexp,yy,z,x(NPT),y(NPT)
      f(z)=z*exp(-z)/((z-1.0)**2+EPSSQ)
      do 11 i=1,NPT
```

```
          x(i)=i*2.0/NPT
          y(i)=f(x(i))
11        continue
          write(*,'(/1x,a/)') 'Diagonal rational function interpolation'
          write(*,'(1x,t6,a,t13,a,t26,a,t40,a)')
     *         'x','interp.','accuracy','actual'
          do 12 i=1,10
            xx=0.2*i
            call ratint(x,y,NPT,xx,yy,dyy)
            yexp=f(xx)
            write(*,'(1x,f6.2,f12.6,e15.4,f12.6)') xx,yy,dyy,yexp
12        continue
          END
```

Subroutine `spline` generates a cubic spline. Given an array of x_i and $f(x_i)$, and given values of the first derivative of function f at the two endpoints of the tabulated region, it returns the second derivative of f at each of the tabulation points. As an example we chose the function $\sin x$ and evaluated it at evenly spaced points `x(i)`. In this case the first derivatives at the end-points are $yp1 = \cos x_1$ and $ypn = \cos x_N$. The output array of `spline` is `y2(i)` and this is listed along with $-\sin x_i$, the second derivative of $\sin x_i$, for comparison.

```
          PROGRAM xspline
C         driver for routine spline
          INTEGER N
          REAL PI
          PARAMETER(N=20,PI=3.141593)
          INTEGER i
          REAL yp1,ypn,x(N),y(N),y2(N)
          write(*,*) 'Second-derivatives for sin(x) from 0 to PI'
C         generate array for interpolation
          do 11 i=1,20
            x(i)=i*PI/N
            y(i)=sin(x(i))
11        continue
C         calculate 2nd derivative with SPLINE
          yp1=cos(x(1))
          ypn=cos(x(N))
          call spline(x,y,N,yp1,ypn,y2)
C         test result
          write(*,'(t19,a,t35,a)') 'spline','actual'
          write(*,'(t6,a,t17,a,t33,a)') 'angle','2nd deriv','2nd deriv'
          do 12 i=1,N
            write(*,'(1x,f8.2,2f16.6)') x(i),y2(i),-sin(x(i))
12        continue
          END
```

Actual cubic-spline interpolations, however, are carried out by `splint`. This routine uses the output array from one call to `spline` to service any subsequent number of spline interpolations with different x's. The demonstration program `xsplint` tests this capability on both $\sin x$ and $\exp x$. The two are treated in succession according to whether `nfunc` is one or two. In each case the function is tabulated at equally spaced points, and the derivatives are found at the first and last point. A call to `spline` then produces an array of second derivatives `y2` which is fed to `splint`. The interpolated values `y` are compared with actual function values `f` at a different set of equally spaced points.

```
        PROGRAM xsplint
C       driver for routine splint, which calls spline
        INTEGER NP
        REAL PI
        PARAMETER(NP=10,PI=3.141593)
        INTEGER i,nfunc
        REAL f,x,y,yp1,ypn,xa(NP),ya(NP),y2(NP)
        do 14 nfunc=1,2
          if (nfunc.eq.1) then
            write(*,*) 'Sine function from 0 to PI'
            do 11 i=1,NP
              xa(i)=i*PI/NP
              ya(i)=sin(xa(i))
11          continue
            yp1=cos(xa(1))
            ypn=cos(xa(NP))
          else if (nfunc.eq.2) then
            write(*,*) 'Exponential function from 0 to 1'
            do 12 i=1,NP
              xa(i)=1.0*i/NP
              ya(i)=exp(xa(i))
12          continue
            yp1=exp(xa(1))
            ypn=exp(xa(NP))
          else
            stop
          endif
C       call SPLINE to get second derivatives
          call spline(xa,ya,NP,yp1,ypn,y2)
C       call SPLINT for interpolations
          write(*,'(1x,t10,a1,t20,a4,t28,a13)') 'x','f(x)','interpolation'
          do 13 i=1,10
            if (nfunc.eq.1) then
              x=(-0.05+i/10.0)*PI
              f=sin(x)
            else if (nfunc.eq.2) then
              x=-0.05+i/10.0
              f=exp(x)
            endif
            call splint(xa,ya,y2,NP,x,y)
            write(*,'(1x,3f12.6)') x,f,y
13        continue
          write(*,*) '*********************************'
          write(*,*) 'Press RETURN'
          read(*,*)
14      continue
        END
```

The next program, `locate`, may be used in conjunction with any interpolation method to bracket the x-position for which $f(x)$ is sought by two adjacent tabulated positions. That is, given a monotonic array of x_i, and given a value of x, it finds the two values x_i, x_{i+1} that surround x. In `xlocate` we chose the array x_i to be non-uniform, varying exponentially with i. Then we took a uniform series of x-values and sought their position in the array using `locate`. For each x, `locate` finds the value j for which xx(j) is nearest below x. Then the driver shows j, and the two bracketing values xx(j) and xx(j+1). If j is 0 or N, then x is not within the tabulated range. The program then flags 'lower lim' if x is below xx(1) or 'upper lim' if x is above xx(N).

```
      PROGRAM xlocate
C     driver for routine locate
      INTEGER N
      PARAMETER(N=100)
      INTEGER i,j
      REAL x,xx(N)
C     create array to be searched
      do 11 i=1,N
        xx(i)=exp(i/20.0)-74.0
11    continue
      write(*,*) 'Result of:   j=0 indicates x too small'
      write(*,*) '             j=100 indicates x too large'
      write(*,'(t5,a7,t17,a1,t24,a5,t34,a7)') 'locate ','j',
     *     'xx(j)','xx(j+1)'
C     perform test
      do 12 i=1,19
        x=-100.0+200.0*i/20.0
        call locate(xx,N,x,j)
        if (j.eq.0) then
          write(*,'(1x,f10.4,i6,a12,f12.6)') x,j,'lower lim',xx(j+1)
        else if (j.eq.N) then
          write(*,'(1x,f10.4,i6,f12.6,a12)') x,j,xx(j),'upper lim'
        else
          write(*,'(1x,f10.4,i6,2f12.6)') x,j,xx(j),xx(j+1)
        endif
12    continue
      END
```

Routine hunt serves the same function as locate, but is used when the table is to be searched many times and the abscissa each time is close to its value on the previous search. xhunt sets up the array xx(i) and then a series x of points to locate. The hunt begins with a trial value ji (which is fed to hunt through variable j) and hunt returns solution j such that x lies between xx(j) and xx(j+1). The two cases j=0 and j=N have the same meaning as in xlocate and are treated in the same way.

```
      PROGRAM xhunt
C     driver for routine hunt
      INTEGER N
      PARAMETER(N=100)
      INTEGER i,j,ji
      REAL x,xx(N)
C     create array to be searched
      do 11 i=1,N
        xx(i)=exp(i/20.0)-74.0
11    continue
      write(*,*) 'Result of:   j=0 indicates x too small'
      write(*,*) '             j=100 indicates x too large'
      write(*,'(t7,a7,t17,a5,t25,a1,t32,a5,t42,a7)') 'locate:',
     *     'guess','j','xx(j)','xx(j+1)'
C     perform test
      do 12 i=1,19
```

```
      x=-100.0+200.0*i/20.0
C     trial parameter
      ji=5*i
      j=ji
C     begin search
      call hunt(xx,N,x,j)
      if (j.eq.0) then
        write(*,'(1x,f12.6,2i6,a12,f12.6)') x,ji,j,
     *        'lower lim',xx(j+1)
      else if (j.eq.N) then
        write(*,'(1x,f12.6,2i6,f12.6,a12)') x,ji,j,
     *        xx(j),'upper lim'
      else
        write(*,'(1x,f12.6,2i6,2f12.6)') x,ji,j,xx(j),xx(j+1)
      endif
12    continue
      END
```

The next two demonstration programs, xpolcoe and xpolcof, are so nearly identical that they may be discussed together. polcoe and polcof themselves both find
coefficients of interpolating polynomials. In the present instance we have tried both
a sine function and an exponential function for ya(i), each tabulated at uniformly
spaced points xa(i). The validity of the array of polynomial coefficients coeff is
tested by calculating the value sum of the polynomials at a series of test points and
listing these alongside the functions f that they represent.

```
      PROGRAM xpolcoe
C     driver for routine polcoe
      INTEGER NP
      REAL PI
      PARAMETER(NP=5,PI=3.1415926)
      INTEGER i,j,nfunc
      REAL f,sum,x,xa(NP),ya(NP),coeff(NP)
      do 15 nfunc=1,2
        if (nfunc.eq.1) then
          write(*,*) 'Sine function from 0 to PI'
          do 11 i=1,NP
            xa(i)=i*PI/NP
            ya(i)=sin(xa(i))
11        continue
        else if (nfunc.eq.2) then
          write(*,*) 'Exponential function from 0 to 1'
          do 12 i=1,NP
            xa(i)=1.0*i/NP
            ya(i)=exp(xa(i))
12        continue
        else
          stop
        endif
        call polcoe(xa,ya,NP,coeff)
        write(*,*) '    coefficients'
        write(*,'(1x,6f12.6)') (coeff(i),i=1,NP)
        write(*,'(1x,t10,a1,t20,a4,t29,a10)')
     *        'x','f(x)','polynomial'
        do 14 i=1,10
          if (nfunc.eq.1) then
```

```
            x=(-0.05+i/10.0)*PI
            f=sin(x)
          else if (nfunc.eq.2) then
            x=-0.05+i/10.0
            f=exp(x)
          endif
          sum=coeff(NP)
          do 13 j=NP-1,1,-1
            sum=coeff(j)+sum*x
13        continue
          write(*,'(1x,3f12.6)') x,f,sum
14      continue
        write(*,*) '********************************'
        write(*,*) 'Press RETURN'
        read(*,*)
15    continue
      END

      PROGRAM xpolcof
C     driver for routine polcof
      INTEGER NP
      REAL PI
      PARAMETER(NP=5,PI=3.141593)
      INTEGER i,j,nfunc
      REAL f,sum,x,xa(NP),ya(NP),coeff(NP)
      do 15 nfunc=1,2
        if (nfunc.eq.1) then
          write(*,*) 'Sine function from 0 to PI'
          do 11 i=1,NP
            xa(i)=i*PI/NP
            ya(i)=sin(xa(i))
11        continue
        else if (nfunc.eq.2) then
          write(*,*) 'Exponential function from 0 to 1'
          do 12 i=1,NP
            xa(i)=1.0*i/NP
            ya(i)=exp(xa(i))
12        continue
        else
          stop
        endif
        call polcof(xa,ya,NP,coeff)
        write(*,*) '   coefficients'
        write(*,'(1x,6f12.6)') (coeff(i),i=1,NP)
        write(*,'(1x,t10,a1,t20,a4,t29,a10)')
     *       'x','f(x)','polynomial'
        do 14 i=1,10
          if (nfunc.eq.1) then
            x=(-0.05+i/10.0)*PI
            f=sin(x)
          else if (nfunc.eq.2) then
            x=-0.05+i/10.0
            f=exp(x)
          endif
          sum=coeff(NP)
          do 13 j=NP-1,1,-1
            sum=coeff(j)+sum*x
```

```
13        continue
          write(*,'(1x,3f12.6)') x,f,sum
14        continue
          write(*,*) '*********************************'
          write(*,*) 'Press RETURN'
          read(*,*)
15        continue
          END
```

For two-dimensional interpolation, polin2 implements a bilinear interpolation. We feed it coordinates x1a,x2a for an $M \times N$ array of gridpoints as well as the function value at each gridpoint. In return it gives the value y of the interpolated function at a given point x1,x2, and the estimated accuracy dy of the interpolation. xpolin2 runs the test on a uniform grid for the function $f(x, y) = \sin x \exp y$. Then, for an offset grid of test points, the interpolated value y is compared to the actual function value f, and the actual error is compared to the estimated error dy.

```
          PROGRAM xpolin2
C         driver for routine polin2
          INTEGER N
          REAL PI
          PARAMETER(N=5,PI=3.141593)
          INTEGER i,j
          REAL dy,f,x1,x2,y,x1a(N),x2a(N),ya(N,N)
          do 12 i=1,N
            x1a(i)=i*PI/N
            do 11 j=1,N
              x2a(j)=1.0*j/N
              ya(i,j)=sin(x1a(i))*exp(x2a(j))
11          continue
12        continue
C         test 2-dimensional interpolation
          write(*,'(t9,a,t21,a,t32,a,t40,a,t58,a)')
     *         'x1','x2','f(x)','interpolated','error'
          do 14 i=1,4
            x1=(-0.1+i/5.0)*PI
            do 13 j=1,4
              x2=-0.1+j/5.0
              f=sin(x1)*exp(x2)
              call polin2(x1a,x2a,ya,N,N,x1,x2,y,dy)
              write(*,'(1x,4f12.6,f14.6)') x1,x2,f,y,dy
13          continue
            write(*,*) '*********************************'
14        continue
          END
```

Bicubic interpolation in two dimensions is carried out with bcucof and bcuint. The first supplies interpolating coefficients within a grid square and the second calculates interpolated values. The calculation provides not only interpolated function values, but also interpolated values of two partial derivatives, all of which are guaranteed to be smooth. To get this, we are required to supply more information than we have needed in previous interpolation routines.

Demonstration program xbcucof works with the function $f(x, y) = xy \exp(-xy)$. The program computes the two first derivatives and the cross derivative of this function at the four corners of a rectangular grid cell, in this case a 2×2 unit square with one corner at the origin. The points are supplied counterclockwise around the cell. d1 and d2 are the dimensions of the cell. A call to bcucof provides the following sixteen coefficients, listed for comparison.

```
Coefficients for bicubic interpolation
 0.000000E+00    0.000000E+00    0.000000E+00    0.000000E+00
 0.000000E+00    0.400000E+01    0.000000E+00    0.000000E+00
 0.000000E+00    0.000000E+00   -0.136556E+02    0.609517E+01
 0.000000E+00    0.000000E+00    0.609517E+01   -0.246149E+01
```

```
      PROGRAM xbcucof
C     driver for routine bcucof
      INTEGER i,j
      REAL d1,d2,ee,x1x2
      REAL c(4,4),y(4),y1(4),y2(4)
      REAL y12(4),x1(4),x2(4)
      DATA x1/0.0,2.0,2.0,0.0/
      DATA x2/0.0,0.0,2.0,2.0/
      d1=x1(2)-x1(1)
      d2=x2(4)-x2(1)
      do 11 i=1,4
        x1x2=x1(i)*x2(i)
        ee=exp(-x1x2)
        y(i)=x1x2*ee
        y1(i)=x2(i)*(1.0-x1x2)*ee
        y2(i)=x1(i)*(1.0-x1x2)*ee
        y12(i)=(1.0-3.0*x1x2+x1x2**2)*ee
11    continue
      call bcucof(y,y1,y2,y12,d1,d2,c)
      write(*,*) 'Coefficients for bicubic interpolation'
      do 12 i=1,4
        write(*,'(1x,4e15.6)') (c(i,j),j=1,4)
12    continue
      END
```

Program xbcuint works with the function $f(x,y) = (xy)^2$, which has derivatives $\partial f/\partial x = 2xy^2$, $\partial f/\partial y = 2yx^2$, and $\partial^2 f/\partial x\partial y = 4xy$. These are supplied to bcuint along with the locations of the grid points. bcuint calls bcucof internally to determine coefficients, and then calculates ansy, ansy1, ansy2, the interpolated values of f, $\partial f/\partial x$ and $\partial f/\partial y$ at the specified test point (x1,x2). These are compared by the demonstration program to expected values for the three quantities, which are called ey, ey1, and ey2. The test points run along the diagonal of the grid square.

```
      PROGRAM xbcuint
C     driver for routine bcuint
      INTEGER i
      REAL ansy,ansy1,ansy2,ey,ey1,ey2
      REAL x1,x1l,x1u,x1x2,x2,x2l,x2u,xxyy
      REAL y(4),y1(4),y2(4),y12(4),xx(4),yy(4)
      DATA xx/0.0,2.0,2.0,0.0/
      DATA yy/0.0,0.0,2.0,2.0/
      x1l=xx(1)
      x1u=xx(2)
      x2l=yy(1)
      x2u=yy(4)
```

```
      do 11 i=1,4
        xxyy=xx(i)*yy(i)
        y(i)=xxyy**2
        y1(i)=2.0*yy(i)*xxyy
        y2(i)=2.0*xx(i)*xxyy
        y12(i)=4.0*xxyy
11    continue
      write(*,'(/1x,t6,a,t14,a,t22,a,t28,a,t38,a,t44,a,t54,a,t60,a/)')
     *    'X1','X2','Y','EXPECT','Y1','EXPECT','Y2','EXPECT'
      do 12 i=1,10
        x1=0.2*i
        x2=x1
        call bcuint(y,y1,y2,y12,x1l,x1u,x2l,x2u,x1,x2,ansy,ansy1,ansy2)
        x1x2=x1*x2
        ey=x1x2**2
        ey1=2.0*x2*x1x2
        ey2=2.0*x1*x1x2
        write(*,'(1x,8f8.4)') x1,x2,ansy,ey,ansy1,ey1,ansy2,ey2
12    continue
      END
```

Routines splie2 and splin2 work as a pair to perform bicubic spline interpolations. splie2 takes a function tabulated on an $M \times N$ grid and performs one dimensional natural cubic splines along the rows of the grid to generate an array of second derivatives. These are fodder for splin2 which takes the grid points, function values, and second derivative values and returns the interpolated function value for a desired point in the grid region.

Demonstration program xsplie2 exercises splie2 on a regular 10×10 grid of points with coordinates x1 and x2, for the function $y = (x_1 x_2)^2$. The calculated second derivative array is compared with the actual second derivative $2x_1 x_2$ of the function. Keep in mind that a natural spline is assumed, so that agreement will not be so good near the boundaries of the grid. (This shows that you should *not* assume a natural spline if you have better derivative information at the endpoints.)

```
      PROGRAM xsplie2
C     driver for routine splie2
      INTEGER M,N
      PARAMETER(M=10,N=10)
      INTEGER i,j
      REAL x1x2,x1(M),x2(N),y(M,N),y2(M,N)
      do 11 i=1,M
        x1(i)=0.2*i
11    continue
      do 12 i=1,N
        x2(i)=0.2*i
12    continue
      do 14 i=1,M
        do 13 j=1,N
          x1x2=x1(i)*x2(j)
          y(i,j)=x1x2**2
13      continue
14    continue
      call splie2(x1,x2,y,M,N,y2)
      write(*,'(/1x,a)') 'Second derivatives from SPLIE2'
      write(*,'(1x,a/)') 'Natural spline assumed'
```

```
        do 15 i=1,5
        write(*,'(1x,5f12.6)') (y2(i,j),j=1,5)
15      continue
        write(*,'(/1x,a/)') 'Actual second derivatives'
        do 17 i=1,5
          do 16 j=1,5
            y2(i,j)=2.0*(x1(i)**2)
16        continue
          write(*,'(1x,5f12.6)') (y2(i,j),j=1,5)
17      continue
        END
```

The demonstration program xsplin2 establishes a similar 10×10 grid for the function $y = x_1 x_2 \exp(-x_1 x_2)$. It makes a single call to splie2 to produce second derivatives y2, and then finds function values f through calls to splin2, comparing them to actual function values ff. These values are determined, for no particular reason, along a quadratic path $x_2 = x_1^2$ through the grid region.

```
        PROGRAM xsplin2
C       driver for routine splin2
        INTEGER M,N
        PARAMETER(M=10,N=10)
        INTEGER i,j
        REAL f,ff,x1x2,xx1,xx2,x1(M),x2(N),y(M,N),y2(M,N)
        do 11 i=1,M
          x1(i)=0.2*i
11      continue
        do 12 i=1,N
          x2(i)=0.2*i
12      continue
        do 14 i=1,M
          do 13 j=1,N
            x1x2=x1(i)*x2(j)
            y(i,j)=x1x2*exp(-x1x2)
13        continue
14      continue
        call splie2(x1,x2,y,M,N,y2)
        write(*,'(/1x,t9,a,t21,a,t31,a,t43,a)')
     *      'x1','x2','splin2','actual'
        do 15 i=1,10
          xx1=0.1*i
          xx2=xx1**2
          call splin2(x1,x2,y,y2,M,N,xx1,xx2,f)
          x1x2=xx1*xx2
          ff=x1x2*exp(-x1x2)
          write(*,'(1x,4f12.6)') xx1,xx2,f,ff
15      continue
        END
```

Chapter 4: Integration of Functions

Numerical integration, or "quadrature", has been treated with some degree of detail in Numerical Recipes. Chapter 4 begins with `trapzd`, *a subroutine for applying the extended trapezoidal rule. It can be used in successive calls for sequentially improving accuracy, and is used as a foundation for several other programs. For example* `qtrap` *is an integrating routine that makes repeated calls to* `trapzd` *until a certain fractional accuracy is achieved.* `qsimp` *also calls* `trapzd`, *and in this case performs integration by Simpson's rule. Romberg integration, a generalization of Simpson's rule to successively higher orders, is performed with* `qromb` — *this one also calls* `trapzd`. *For improper integrals a different "workhorse" is used, the subroutine* `midpnt`. *This routine applies the extended midpoint rule to avoid function evaluations at an endpoint of the region of integration. It can be used in* `qtrap` *or* `qsimp` *in place of* `trapzd`. *Routine* `qromb` *can be generalized similarly, and we have implemented this idea in* `qromo`, *a Romberg integrator for open intervals. The chapter also offers a number of exact replacements for* `midpnt`, *to be used for various types of singularity in the integrand:*

1. `midinf` – *if one or the other of the limits of integration is infinite.*

2. `midsql` – *if there is an inverse square root singularity of the integrand at the lower limit of integration.*

3. `midsqu` – *if there is an inverse square root singularity of the integrand at the upper limit of integration.*

4. `midexp` – *when the upper limit of integration is infinite and the integrand decreases exponentially at infinity.*

The somewhat more subtle method of Gaussian quadrature uses unequally spaced abscissas, and weighting coefficients which can be read from tables. Routine `qgaus` *computes integrals with a ten-point Gauss-Legendre weighting using such coefficients.* `gauleg, gaulag, gauher` *and* `gaujac` *calculate the tables of abscissas and weights for N-point Gauss-Legendre, Gauss-Laguerre, Gauss-Hermite and Gauss-Jacobi quadratures, respectively. The routine* `orthog` *lets you construct nonclassical orthogonal polynomials, while* `gaucof` *lets you find the abscissas and weights for the corresponding Gaussian quadrature.*

$\star$ $\quad$ $\star$ $\quad$ $\star$ $\quad$ $\star$

`trapzd` applies the extended trapezoidal rule for integration. It is called sequentially for higher and higher stages of refinement of the integral. The sample program `xtrapzd` uses `trapzd` to perform a numerical integration of the function

$$\text{func} = x^2(x^2 - 2)\sin x$$

whose indefinite integral is

$$\text{fint} = 4x(x^2 - 7)\sin x - (x^4 - 14x^2 + 28)\cos x.$$

The integral is performed from $A = 0.0$ to $B = \pi/2$. To demonstrate the increasing accuracy on sequential calls, `trapzd` is called 14 times with the index i increasing by one each time. The improving values of the integral are listed for comparison to the actual value $\text{fint}(B) - \text{fint}(A)$.

```
      PROGRAM xtrapzd
C     driver for routine trapzd
      INTEGER NMAX
      REAL PIO2
      PARAMETER(NMAX=14, PIO2=1.5707963)
      INTEGER i
      REAL a,b,fint,func,s
      EXTERNAL func
      a=0.0
      b=PIO2
      write(*,'(1x,a)') 'Integral of FUNC with 2^(n-1) points'
      write(*,'(1x,a,f10.6)') 'Actual value of integral is',
     *    fint(b)-fint(a)
      write(*,'(1x,t7,a,t16,a)') 'n','Approx. Integral'
      do 11 i=1,NMAX
        call trapzd(func,a,b,s,i)
        write(*,'(1x,i6,f20.6)') i,s
11    continue
      END

      REAL FUNCTION func(x)
      REAL x
      func=(x**2)*(x**2-2.0)*sin(x)
      END

      REAL FUNCTION fint(x)
C     integral of FUNC
      REAL x
      fint=4.0*x*((x**2)-7.0)*sin(x)-((x**4)-14.0*(x**2)+28.0)*cos(x)
      END
```

`qtrap` carries out the same integration algorithm but allows us to specify the accuracy with which we wish the integration done. (It is specified within `qtrap` as `EPS=1.0E-6`.) `qtrap` itself makes the sequential calls to `trapzd` until the desired accuracy is reached. Then `qtrap` issues a single result. In sample program `xqtrap` we compare this result to the exact value of the integral.

```
      PROGRAM xqtrap
C     driver for routine qtrap
      REAL PIO2
      PARAMETER(PIO2=1.5707963)
      REAL a,b,fint,func,s
      EXTERNAL func
      a=0.0
```

```
      b=PIO2
      write(*,'(1x,a)') 'Integral of FUNC computed with QTRAP'
      write(*,'(1x,a,f10.6)') 'Actual value of integral is',
     *    fint(b)-fint(a)
      call qtrap(func,a,b,s)
      write(*,'(1x,a,f10.6)') 'Result from routine QTRAP is',s
      END

      REAL FUNCTION func(x)
      REAL x
      func=(x**2)*(x**2-2.0)*sin(x)
      END

      REAL FUNCTION fint(x)
C     integral of FUNC
      REAL x
      fint=4.0*x*((x**2)-7.0)*sin(x)-((x**4)-14.0*(x**2)+28.0)*cos(x)
      END
```

Alternatively, the integral may be handled by `qsimp` which applies Simpson's rule. Sample program `xqsimp` carries out the same integration as the previous program, and reports the result in the same way as well.

```
      PROGRAM xqsimp
C     driver for routine qsimp
      REAL PIO2
      PARAMETER(PIO2=1.5707963)
      REAL a,b,fint,func,s
      EXTERNAL fint,func
      a=0.0
      b=PIO2
      write(*,'(1x,a)') 'Integral of FUNC computed with QSIMP'
      write(*,'(1x,a,f10.6)') 'Actual value of integral is',
     *    fint(b)-fint(a)
      call qsimp(func,a,b,s)
      write(*,'(1x,a,f10.6)') 'Result from routine QSIMP is',s
      END

      REAL FUNCTION func(x)
      REAL x
      func=(x**2)*(x**2-2.0)*sin(x)
      END

      REAL FUNCTION fint(x)
C     integral of FUNC
      REAL x
      fint=4.0*x*((x**2)-7.0)*sin(x)-((x**4)-14.0*(x**2)+28.0)*cos(x)
      END
```

`qromb` generalizes Simpson's rule to higher orders. It makes successive calls to `trapzd` and stores the results. Then it uses `polint`, the polynomial interpolater/extrapolator, to project the value of integral that would be obtained were we to continue indefinitely with `trapzd`. Sample program `xqromb` is essentially identical to the sample programs for `qtrap` and `qsimp`.

```
      PROGRAM xqromb
C     driver for routine qromb
      REAL PIO2
      PARAMETER(PIO2=1.5707963)
      REAL a,b,fint,func,s
      EXTERNAL fint,func
      a=0.0
      b=PIO2
      write(*,'(1x,a)') 'Integral of FUNC computed with QROMB'
      write(*,'(1x,a,f10.6)') 'Actual value of integral is',
     *     fint(b)-fint(a)
      call qromb(func,a,b,s)
      write(*,'(1x,a,f10.6)') 'Result from routine QROMB is',s
      END

      REAL FUNCTION func(x)
      REAL x
      func=(x**2)*(x**2-2.0)*sin(x)
      END

      REAL FUNCTION fint(x)
C     integral of FUNC
      REAL x
      fint=4.0*x*((x**2)-7.0)*sin(x)-((x**4)-14.0*(x**2)+28.0)*cos(x)
      END
```

Sample program `xmidpnt` uses the function `func2` $= 1/\sqrt{x}$ which is singular at the origin. Limits of integration are set at $A = 0.0$ and $B = 1.0$. `midpnt`, however, implements an open formula and does not evaluate the function exactly at $x = 0$. In this case the integral is compared to `fint2`$(B) -$ `fint2`(A) where `fint2` $= 2\sqrt{x}$, the integral of `func2`.

```
      PROGRAM xmidpnt
C     driver for routine midpnt
      INTEGER NMAX
      PARAMETER(NMAX=10)
      INTEGER i
      REAL a,b,fint2,func2,s
      EXTERNAL fint2,func2
      a=0.0
      b=1.0
      write(*,*) 'Integral of FUNC2 computed with MIDPNT'
      write(*,*) 'Actual value of integral is',fint2(b)-fint2(a)
      write(*,'(1x,t7,a,t20,a)') 'n','Approx. Integral'
      do 11 i=1,NMAX
        call midpnt(func2,a,b,s,i)
        write(*,'(1x,i6,f24.6)') i,s
11    continue
      END

      REAL FUNCTION func2(x)
      REAL x
      func2=1.0/sqrt(x)
      END

      REAL FUNCTION fint2(x)
C     integral of FUNC2
```

```
REAL x
fint2=2.0*sqrt(x)
END
```

Various special forms of midpnt (i.e., midsql, midsqu, midinf, midexp) are demonstrated by sample program xqromo. For those tests that integrate to infinity, we take infinity to be 1.0E20. The following integrations are performed:

1. Integral of $\sqrt{x}/\sin x$ from 0.0 to $\pi/2$. (This has a $1/\sqrt{x}$ singularity at $x = 0$, and uses midsql.)

2. Integral of $\sqrt{(\pi - x)}/\sin x$ from $\pi/2$ to π. (This has the $1/\sqrt{x}$ singularity at the upper limit $x = \pi$, and uses midsqu.)

3. Integral of $(\sin x)/x^2$ from $\pi/2$ to ∞. (This has a region of integration extending to ∞ and uses midinf. It is quite slowly convergent, as is the next integral.)

4. Integral of $(\sin x)/x^2$ from $-\infty$ to $-\pi/2$. (Region of integration goes to $-\infty$; uses midinf.)

5. Integral of $\exp(-x)/\sqrt{x}$ from 0.0 to ∞. (This has a singularity at $x = 0.0$ and also integrates up to ∞. It is performed in two pieces, (0.0 to $\pi/2$) and ($\pi/2$ to ∞) using midsql and midinf. The two calculations give results res1 and res2 respectively, which are added to give the entire integral.)

6. Same integral as in (5), but with the segment from ($\pi/2$ to ∞) done using midexp.

```
      PROGRAM xqromo
C     driver for routine qromo
      REAL AINF,X1,X2,X3
      PARAMETER(X1=0.0,X2=1.5707963,X3=3.1415926,AINF=1.0E20)
      REAL res1,res2,result
      EXTERNAL funcl,midsql,funcu,midsqu,fncinf,midinf,fncend,midexp
      write(*,'(/1x,a)') 'Improper integrals:'
      call qromo(funcl,X1,X2,result,midsql)
      write(*,'(/1x,a)')
     *     'Function: SQRT(x)/SIN(x)      Interval: (0,pi/2)'
      write(*,'(1x,a,f8.4)')
     *     'Using: MIDSQL               Result:',result
      call qromo(funcu,X2,X3,result,midsqu)
      write(*,'(/1x,a)')
     *     'Function: SQRT(pi-x)/SIN(x)   Interval: (pi/2,pi)'
      write(*,'(1x,a,f8.4)')
     *     'Using: MIDSQU               Result:',result
      call qromo(fncinf,X2,AINF,result,midinf)
      write(*,'(/1x,a)')
     *     'Function: SIN(x)/x**2        Interval: (pi/2,infty)'
      write(*,'(1x,a,f8.4)')
     *     'Using: MIDINF               Result:',result
      call qromo(fncinf,-AINF,-X2,result,midinf)
      write(*,'(/1x,a)')
     *     'Function: SIN(x)/x**2        Interval: (-infty,-pi/2)'
      write(*,'(1x,a,f8.4)')
     *     'Using: MIDINF               Result:',result
      call qromo(fncend,X1,X2,res1,midsql)
      call qromo(fncend,X2,AINF,res2,midinf)
      write(*,'(/1x,a)')
```

```
*       'Function: EXP(-x)/SQRT(x)      Interval: (0.0,infty)'
        write(*,'(1x,a,f8.4)')
*       'Using: MIDSQL,MIDINF           Result:',res1+res2
        call qromo(fncend,X2,AINF,res2,midexp)
        write(*,'(/1x,a)')
*       'Function: EXP(-x)/SQRT(x)      Interval: (0.0,infty)'
        write(*,'(1x,a,f8.4/)')
*       'Using: MIDSQL,MIDEXP           Result:',res1+res2
        END

        REAL FUNCTION funcl(x)
        REAL x
        funcl=sqrt(x)/sin(x)
        END

        REAL FUNCTION funcu(x)
        REAL PI
        PARAMETER(PI=3.1415926)
        REAL x
        funcu=sqrt(PI-x)/sin(x)
        END

        REAL FUNCTION fncinf(x)
        REAL x
        fncinf=sin(dble(x))/(x**2)
        END

        REAL FUNCTION fncend(x)
        REAL x
        fncend=exp(-x)/sqrt(x)
        END
```

Subroutine qgaus performs a Gauss-Legendre integration, using only ten function evaluations. Sample program xqgaus applies it to the function $x\exp(-x)$ whose integral from x_1 to x is $(1 + x_1)\exp(-x_1) - (1 + x)\exp(-x)$. qgaus returns this integral as parameter ss. The method is used for a series of intervals, as short as $(0.0 - 0.5)$ and as long as $(0.0 - 5.0)$. You may observe how the accuracy depends on the interval.

```
        PROGRAM xqgaus
C       driver for routine qgaus
        INTEGER NVAL
        REAL X1,X2
        PARAMETER(X1=0.0,X2=5.0,NVAL=10)
        INTEGER i
        REAL dx,func,ss,x
        EXTERNAL func
        dx=(X2-X1)/NVAL
        write(*,'(/1x,a,t12,a,t23,a/)') '0.0 to','QGAUS','Expected'
        do 11 i=1,NVAL
          x=X1+i*dx
          call qgaus(func,X1,x,ss)
          write(*,'(1x,f5.2,2f12.6)') x,ss,
*         -(1.0+x)*exp(-x)+(1.0+X1)*exp(-X1)
11      continue
        END
```

```
REAL FUNCTION func(x)
REAL x
func=x*exp(-x)
END
```

Sample program `xgauleg`, which drives `gauleg`, performs the same method of quadrature, and on the same function. However, it chooses its own abscissas and weights for the Gauss-Legendre calculation, and is not restricted to a ten-point formula; it can do an N-point calculation for any N. The N abscissas and weights appropriate to an interval $x = 0.0$ to 1.0 are found by sample program `xgauleg` for the case $N = 10$. The results you should find are listed below. Next the program applies these values to a quadrature and compares the result to that from a formal integration.

#	X(I)	W(I)
1	0.013047	0.033336
2	0.067468	0.074726
3	0.160295	0.109543
4	0.283302	0.134633
5	0.425563	0.147762
6	0.574437	0.147762
7	0.716698	0.134633
8	0.839705	0.109543
9	0.932532	0.074726
10	0.986953	0.033336

```
      PROGRAM xgauleg
C     driver for routine gauleg
      INTEGER NPOINT
      REAL X1,X2,X3
      PARAMETER(NPOINT=10,X1=0.0,X2=1.0,X3=10.0)
      INTEGER i
      REAL func,xx,x(NPOINT),w(NPOINT)
      call gauleg(X1,X2,x,w,NPOINT)
      write(*,'(/1x,t3,a,t10,a,t22,a/)') '#','X(I)','W(I)'
      do 11 i=1,NPOINT
        write(*,'(1x,i2,2f12.6)') i,x(i),w(i)
11    continue
C     demonstrate the use of GAULEG for an integral
      call gauleg(X1,X3,x,w,NPOINT)
      xx=0.0
      do 12 i=1,NPOINT
        xx=xx+w(i)*func(x(i))
12    continue
      write(*,'(/1x,a,f12.6)') 'Integral from GAULEG:',xx
      write(*,'(1x,a,f12.6)') 'Actual value:',1.0-(1.0+X3)*exp(-X3)
      END

      REAL FUNCTION func(x)
      REAL x
      func=x*exp(-x)
      END
```

The next three sample programs are all similar to the previous one. They compute abscissas and weights for some form of Gaussian quadrature and print them out. As a check, each routine sums the weights and compares the sum with an analytically known value. If the abscissas are not symmetric about the origin, their sum is also nonzero and can be compared with an analytic value. Next, each program calculates an appropriate integral using the coefficients just determined.

The first such sample program, xgaulag, exercises gaulag by computing coefficients for an N-point Gauss-Laguerre quadrature. The program fixes the parameter $\alpha = 1$. It then calculates the integral

$$\int_0^\infty e^{-x} J_0(x) x \, dx = 2^{-3/2}$$

and compares the answer with the analytic result. As you will find, N of order a dozen is enough to determine the value to full single precision.

```
      PROGRAM xgaulag
C     driver for routine gaulag
      INTEGER NP
      PARAMETER(NP=64)
      INTEGER i,n
      REAL func,alf,checkw,checkx,gammln,xx,x(NP),w(NP)
      alf=1.
1     write(*,*) 'Enter N'
      read(*,*,END=99) n
      call gaulag(x,w,n,alf)
      write(*,'(/1x,t3,a,t10,a,t22,a/)') '#','X(I)','W(I)'
      do 11 i=1,n
        write(*,'(1x,i2,2e14.6)') i,x(i),w(i)
11    continue
      checkx=0.
      checkw=0.
      do 12 i=1,n
        checkx=checkx+x(i)
        checkw=checkw+w(i)
12    continue
      write(*,'(/1x,a,e15.7,a,e15.7)') 'Check value:',checkx,
     *    '  should be:',n*(n+alf)
      write(*,'(/1x,a,e15.7,a,e15.7)') 'Check value:',checkw,
     *    '  should be:',exp(gammln(1.+alf))
C     demonstrate the use of GAULAG for an integral
      xx=0.0
      do 13 i=1,n
        xx=xx+w(i)*func(x(i))
13    continue
      write(*,'(/1x,a,f12.6)') 'Integral from GAULAG:',xx
      write(*,'(1x,a,f12.6)') 'Actual value:          ',1./(2.*sqrt(2.))
      go to 1
99    stop
      END

      REAL FUNCTION func(x)
      REAL x,bessj0
      func=bessj0(x)
      END
```

The next demonstration program, xgauher, computes coefficients for Gauss-

Hermite quadrature by calling gauher. The test integral in this case is

$$\int_{-\infty}^{\infty} e^{-x^2} \cos x \, dx = \sqrt{\pi} e^{-1/4}$$

```
      PROGRAM xgauher
C     driver for routine gauher
      INTEGER NP
      REAL SQRTPI
      PARAMETER(NP=64,SQRTPI=1.7724539)
      INTEGER i,n
      REAL func,check,xx,x(NP),w(NP)
1     write(*,*) 'Enter N'
      read(*,*,END=99) n
      call gauher(x,w,n)
      write(*,'(/1x,t3,a,t10,a,t22,a/)') '#','X(I)','W(I)'
      do 11 i=1,n
        write(*,'(1x,i2,2e14.6)') i,x(i),w(i)
11    continue
      check=0.
      do 12 i=1,n
        check=check+w(i)
12    continue
      write(*,'(/1x,a,e15.7,a,e15.7)') 'Check value:',check,
     *   ' should be:',SQRTPI
C     demonstrate the use of GAUHER for an integral
      xx=0.0
      do 13 i=1,n
        xx=xx+w(i)*func(x(i))
13    continue
      write(*,'(/1x,a,f12.6)') 'Integral from GAUHER:',xx
      write(*,'(1x,a,f12.6)') 'Actual value:      ',SQRTPI*exp(-.25)
      go to 1
99    stop
      END

      REAL FUNCTION func(x)
      REAL x
      func=cos(x)
      END
```

Program xgaujac exercises gaujac by computing the abscissas and weights for an N-point Gauss-Jacobi quadrature. A convenient test integral is

$$\int_{-1}^{1} \frac{dx}{\sqrt{1 - x^2}} \frac{1}{\sqrt{1 - k^2(1 + x)/2}} = 2K(k)$$

so we set $\alpha = \beta = -1/2$ in the program. (The integration reduces to Gauss-Chebyshev quadrature for this choice of parameters.) We choose $k = 1/2$, but you can easily experiment with other values. A surprisingly small value of N will calculate the integral accurately.

```
      PROGRAM xgaujac
C     driver for routine gaujac
      INTEGER NP
      REAL PIBY2
      PARAMETER(NP=64,PIBY2=1.5707963)
      INTEGER i,n
      REAL func,ak,alf,bet,checkw,checkx,ellf,gammln,xx,x(NP),w(NP)
      alf=-.5
      bet=-.5
1     write(*,*) 'Enter N'
      read(*,*,END=99) n
      call gaujac(x,w,n,alf,bet)
      write(*,'(/1x,t3,a,t10,a,t22,a/)') '#','X(I)','W(I)'
      do 11 i=1,n
        write(*,'(1x,i2,2e14.6)') i,x(i),w(i)
11    continue
      checkx=0.
      checkw=0.
      do 12 i=1,n
        checkx=checkx+x(i)
        checkw=checkw+w(i)
12    continue
      write(*,'(/1x,a,e15.7,a,e15.7)') 'Check value:',checkx,
     *  ' should be:',n*(bet-alf)/(alf+bet+2*n)
      write(*,'(/1x,a,e15.7,a,e15.7)') 'Check value:',checkw,
     *  ' should be:',exp(gammln(1.+alf)+gammln(1.+bet)-
     *  gammln(2.+alf+bet))*2**(alf+bet+1.)
C     demonstrate the use of GAUJAC for an integral
      ak=.5
      xx=0.0
      do 13 i=1,n
        xx=xx+w(i)*func(ak,x(i))
13    continue
      write(*,'(/1x,a,f12.6)') 'Integral from GAUJAC:',xx
      write(*,'(1x,a,f12.6)') 'Actual value:        ',2.*ellf(PIBY2,ak)
      go to 1
99    stop
      END

      REAL FUNCTION func(ak,x)
      REAL ak,x
      func=1./sqrt(1.-ak**2*(1.+x)/2.)
      END
```

Routine gaucof computes abscissas and weights given the coefficients in the recurrence relation for the corresponding orthogonal polynomials. We test it in xgaucof with Gauss-Hermite quadrature. The results should be the same as from xgauher.

```
      PROGRAM xgaucof
C     driver for routine gaucof
      INTEGER NP
      REAL SQRTPI
      PARAMETER(NP=64,SQRTPI=1.7724539)
      INTEGER i,n
      REAL amu0,check,a(NP),b(NP),x(NP),w(NP)
1     write(*,*) 'Enter N'
      read(*,*,END=99) n
```

```
      do 11 i=1,n-1
        a(i)=0.
        b(i+1)=i*.5
11    continue
      a(n)=0.
C     b(1) is arbitrary for call to TQLI
      amu0=SQRTPI
      call gaucof(n,a,b,amu0,x,w)
      write(*,'(/1x,t3,a,t10,a,t22,a/)') '#','X(I)','W(I)'
      do 12 i=1,n
        write(*,'(1x,i2,2e14.6)') i,x(i),w(i)
12    continue
      check=0.
      do 13 i=1,n
        check=check+w(i)
13    continue
      write(*,'(/1x,a,e15.7,a,e15.7)') 'Check value:',check,
     *  '  should be:',SQRTPI
      go to 1
99    stop
      END
```

Routine `orthog` computes abscissas and weights for a nonclassical weight function. We test it in `xorthog` with the weight function $-\log x$, as described in *Numerical Recipes*. The test integral here is

$$-\int_0^1 \frac{\log x}{(1+x)^2}\, dx = \log 2$$

Once again a rather small value of N suffices for good accuracy.

```
      PROGRAM xorthog
C     driver for routine orthog
      INTEGER NP
      PARAMETER(NP=64)
      INTEGER i,n
      REAL func,amu0,check,xx,a(NP),b(NP),x(NP),w(NP),anu(2*NP),
     *    alpha(2*NP-1),beta(2*NP-1)
1     write(*,*) 'Enter N'
      read(*,*,END=99) n
      alpha(1)=.5
      beta(1)=1.
      do 11 i=2,2*n-1
        alpha(i)=0.5
        beta(i)=1./(4.*(4.-1./(i-1)**2))
11    continue
      anu(1)=1.
      anu(2)=-.25
      do 12 i=2,2*n-1
        anu(i+1)=-anu(i)*i*(i-1)/(2.*(i+1)*(2*i-1))
12    continue
      call orthog(n,anu,alpha,beta,a,b)
      amu0=1.
      call gaucof(n,a,b,amu0,x,w)
      write(*,'(/1x,t3,a,t10,a,t22,a/)') '#','X(I)','W(I)'
      do 13 i=1,n
        write(*,'(1x,i2,2e14.6)') i,x(i),w(i)
```

```
13    continue
      check=0.
      do 14 i=1,n
        check=check+w(i)
14    continue
      write(*,'(/1x,a,e15.7,a,e15.7)') 'Check value:',check,
     *   '  should be:',amu0
C     demonstrate the use of ORTHOG for an integral
      xx=0.0
      do 15 i=1,n
        xx=xx+w(i)*func(x(i))
15    continue
      write(*,'(/1x,a,f12.6)') 'Integral from ORTHOG:',xx
      write(*,'(1x,a,f12.6)') 'Actual value:        ',log(2.)
      go to 1
99    stop
      END


      REAL FUNCTION func(x)
      REAL x
      func=1./(1.+x)**2
      END
```

Chapter 4 of *Numerical Recipes* ends with a short discussion of multidimensional integration, exemplified by routine `quad3d` which does a 3-dimensional integration by repeated 1-dimensional integration. Sample program `xquad3d` applies the method to the integration of `func` $= x^2 + y^2 + z^2$ over a spherical volume with a radius `xmax` which is taken successively as $0.1, 0.2, \ldots, 1.0$. The integral is done in Cartesian rather than spherical coordinates, but the result is compared to that easily found in spherical coordinates, $4\pi(\texttt{xmax})^5/5$. The concept is quite simple, but you may find the zoo of subroutines somewhat confusing. Subroutine `func` generates the function. Subroutines `y1` and `y2` supply the two limits of the y-integration for each value of x. Similarly `z1` and `z2` give the limits of z-integration for given x and y. Routines `qgausx`, `qgausy`, and `qgausz` are all identical except for their names, and are used to do Gauss-Legendre integration along the three coordinate directions.

```
      PROGRAM xquad3d
C     driver for routine quad3d
      COMMON xmax
      REAL xmax
      INTEGER NVAL
      REAL PI
      PARAMETER(PI=3.1415926,NVAL=10)
      INTEGER i
      REAL s,xmin
      write(*,*) 'Integral of r^2 over a spherical volume'
      write(*,'(/4x,a,t14,a,t24,a)') 'Radius','QUAD3D','Actual'
      do 11 i=1,NVAL
        xmax=0.1*i
        xmin=-xmax
        call quad3d(xmin,xmax,s)
        write(*,'(1x,f8.2,2f10.4)') xmax,s,4.0*PI*(xmax**5)/5.0
11    continue
      END
```

```
REAL FUNCTION func(x,y,z)
REAL x,y,z
func=x**2+y**2+z**2
END

REAL FUNCTION z1(x,y)
REAL x,y
COMMON xmax
REAL xmax
z1=-sqrt(abs(xmax**2-x**2-y**2))
END

REAL FUNCTION z2(x,y)
REAL x,y
COMMON xmax
REAL xmax
z2=sqrt(abs(xmax**2-x**2-y**2))
END

REAL FUNCTION y1(x)
REAL x
COMMON xmax
REAL xmax
y1=-sqrt(abs(xmax**2-x**2))
END

REAL FUNCTION y2(x)
REAL x
COMMON xmax
REAL xmax
y2=sqrt(abs(xmax**2-x**2))
END

SUBROUTINE qgausx(func,a,b,ss)
REAL x(5),w(5)
INTEGER j
REAL a,b,dx,ss,xm,xr
REAL func
EXTERNAL func
DATA x/.1488743389,.4333953941,.6794095682,
*      .8650633666,.9739065285/
DATA w/.2955242247,.2692667193,.2190863625,
*      .1494513491,.0666713443/
xm=0.5*(b+a)
xr=0.5*(b-a)
ss=0
do 11 j=1,5
  dx=xr*x(j)
  ss=ss+w(j)*(func(xm+dx)+func(xm-dx))
11   continue
ss=xr*ss
return
END

SUBROUTINE qgausy(func,a,b,ss)
REAL x(5),w(5)
INTEGER j
```

```
      REAL a,b,dx,ss,xm,xr
      REAL func
      EXTERNAL func
      DATA x/.1488743389,.4333953941,.6794095682,
     *    .8650633666,.9739065285/
      DATA w/.2955242247,.2692667193,.2190863625,
     *    .1494513491,.0666713443/
      xm=0.5*(b+a)
      xr=0.5*(b-a)
      ss=0
      do 11 j=1,5
        dx=xr*x(j)
        ss=ss+w(j)*(func(xm+dx)+func(xm-dx))
11    continue
      ss=xr*ss
      return
      END


      SUBROUTINE qgausz(func,a,b,ss)
      REAL x(5),w(5)
      INTEGER j
      REAL a,b,dx,ss,xm,xr
      REAL func
      EXTERNAL func
      DATA x/.1488743389,.4333953941,.6794095682,
     *    .8650633666,.9739065285/
      DATA w/.2955242247,.2692667193,.2190863625,
     *    .1494513491,.0666713443/
      xm=0.5*(b+a)
      xr=0.5*(b-a)
      ss=0
      do 11 j=1,5
        dx=xr*x(j)
        ss=ss+w(j)*(func(xm+dx)+func(xm-dx))
11    continue
      ss=xr*ss
      return
      END
```

Chapter 5: Evaluation of Functions

Chapter 5 of *Numerical Recipes* treats the approximation and evaluation of functions. The methods, along with a few others, are applied in Chapter 6 to the calculation of a collection of "special" functions. Polynomial or power series expansions are perhaps the most often used approximations and a few tips are given for accelerating the convergence of some series. In the case of alternating series, Euler's transformation is popular, and is implemented in program `eulsum`. For general polynomials, `ddpoly` demonstrates the evaluation of both the polynomial and its derivatives from a list of its coefficients. The division of one polynomial into another, giving a quotient and remainder polynomial, is done by `poldiv`, while the evaluation of a rational function as the ratio of two polynomials is carried out by `ratval`. Ridders' method of finding numerical derivatives, `dfridr`, relies on polynomial extrapolation techniques from Chapter 3.

The approximation of functions by Chebyshev polynomial series is presented as a method of arriving at the approximation of nearly smallest deviation from the true function over a given region for a specified order of approximation. The coefficients for such polynomials are given by `chebft` and function approximations are subsequently carried out by `chebev`. To generate the derivative or integral of a function from its Chebyshev coefficients, use `chder` or `chint` respectively. To convert Chebyshev coefficients into coefficients of a polynomial for the same function (a dangerous procedure about which we offer due warning in the text) use `chebpc` and `pcshft` in succession. Finally, to economize power series, use `pccheb`, the inverse of `chebpc`.

The chapter concludes with a routine for finding the Padé approximant from power series coefficients, `pade`, and a routine `ratlsq` for finding a rational function fit by a least squares method.

Chapter 5 also treats several methods for which we supply no programs. These are continued fractions, complex arithmetic, recurrence relations, and the solution of quadratic and cubic equations.

$$\star \quad \star \quad \star \quad \star$$

Subroutine `eulsum` applies Euler's transformation to the summation of an alternating series. It is called successively for each term to be summed. Our sample program `xeulsum` evaluates the approximation

$$\ln(1+x) = x - \frac{x^2}{2} + \frac{x^3}{3} - \frac{x^4}{4} + \cdots \qquad -1 < x < 1$$

It asks how many terms mval are to be included in the approximation and then makes mval calls to eulsum. Each time, index j increases and term takes the value $(-1)^{j+1}x^j/j$. Both this approximation and the function $\ln(1+x)$ itself are evaluated across the region -1 to 1 for comparison. If mval is set less than 1 or more than 40, the program terminates.

```
      PROGRAM xeulsum
C     driver for routine eulsum
      INTEGER NVAL
      PARAMETER (NVAL=40)
      INTEGER i,j,mval
      REAL sum,term,x,xpower,wksp(NVAL)
C     evaluate ln(1+x)=x-x^2/2+x^3/3-x^4/4... for -1<x<1
10    write(*,*) 'How many terms in polynomial?'
      write(*,'(1x,a,i2,a)') 'Enter n between 1 and ',NVAL,
     *    '. Enter n=0 to END.'
      read(*,*) mval
      if ((mval.le.0).or.(mval.gt.NVAL)) stop
      write(*,'(1x,t9,a1,t18,a6,t28,a10)') 'X','Actual','Polynomial'
      do 12 i=-8,8,1
        x=i/10.0
        sum=0.0
        xpower=-1
        do 11 j=1,mval
          xpower=-x*xpower
          term=xpower/j
          call eulsum(sum,term,j,wksp)
11      continue
        write(*,'(3f12.6)') x,log(1.0+x),sum
12    continue
      goto 10
      END
```

ddpoly evaluates a polynomial and its derivatives, given the coefficients of the polynomial in the form of an input vector. Sample program xddpoly illustrates this for the polynomial:

$$(x-1)^5 = -1 + 5x - 10x^2 + 10x^3 - 5x^4 + x^5$$

(This is a foolish example, of course. No one would knowingly evaluate $(x-1)^5$ by multiplying it out and evaluating terms individually — but it gives us a convenient way to check the result!). Since this is a fifth order polynomial, we set NC, the number of coeffients, to 6, and initialize the array c of coefficients, with c(1) being the constant coefficient and c(6) the highest-order coefficient. There are two loops, one of which evaluates for x values from 0.0 to 2.0, and the other of which stores the value of the function and NC-1 derivatives. d(j,i) keeps the entire array of values for printing. In the second part of the program, the polynomial evaluations are compared with

$$f^{(n-1)}(x) = \frac{5!}{(6-n)!}(x-1.0)^{6-n} \qquad n = 1,\ldots,5$$

```
      PROGRAM xddpoly
C     driver for routine ddpoly
      INTEGER NC,NCM1,NP
      PARAMETER(NC=6,NCM1=5,NP=20)
      INTEGER i,j
      REAL x,factrl,c(NC),pd(NCM1),d(NCM1,NP)
      CHARACTER a(NCM1)*15
      DATA a/'polynomial:','first deriv:','second deriv:',
     *     'third deriv:','fourth deriv:'/
      DATA c/-1.0,5.0,-10.0,10.0,-5.0,1.0/
      do 12 i=1,NP
        x=0.1*i
        call ddpoly(c,NC,x,pd,NC-1)
        do 11 j=1,NC-1
          d(j,i)=pd(j)
11      continue
12    continue
      do 14 i=1,NC-1
        write(*,'(1x,t7,a)') a(i)
        write(*,'(1x,t13,a,t25,a,t40,a)') 'X','DDPOLY','actual'
        do 13 j=1,NP
          x=0.1*j
          write(*,'(1x,3f15.6)') x,d(i,j),
     *          factrl(NC-1)/factrl(NC-i)*((x-1.0)**(NC-i))
13      continue
        write(*,*) 'press ENTER to continue...'
        read(*,*)
14    continue
      END
```

poldiv divides polynomials. Given the coefficients of a numerator and denominator polynomial, poldiv returns the coefficients of a quotient and a remainder polynomial. Sample program xpoldiv takes

$$\text{Numerator} = u = -1 + 5x - 10x^2 + 10x^3 - 5x^4 + x^5 = (x-1)^5$$
$$\text{Denominator} = v = 1 + 3x + 3x^2 + x^3 = (x+1)^3$$

for which we expect

$$\text{Quotient} = q = 31 - 8x + x^2$$
$$\text{Remainder} = r = -32 - 80x - 80x^2$$

The program compares these with the output of poldiv.

```
      PROGRAM xpoldiv
C     driver for routine poldiv
      INTEGER N,NV
      PARAMETER(N=6,NV=4)
      INTEGER i
      REAL u(N),v(NV),q(N),r(N)
      DATA u/-1.0,5.0,-10.0,10.0,-5.0,1.0/
      DATA v/1.0,3.0,3.0,1.0/
      call poldiv(u,N,v,NV,q,r)
      write(*,'(//1x,6(7x,a)/)') 'X^0','X^1','X^2','X^3','X^4','X^5'
      write(*,*) 'Quotient polynomial coefficients:'
      write(*,'(1x,6f10.2/)') (q(i),i=1,6)
```

```
write(*,*) 'Expected quotient coefficients:'
write(*,'(1x,6f10.2///)') 31.0,-8.0,1.0,0.0,0.0,0.0
write(*,*) 'Remainder polynomial coefficients:'
write(*,'(1x,4f10.2/)') (r(i),i=1,4)
write(*,*) 'Expected remainder coefficients:'
write(*,'(1x,4f10.2//)') -32.0,-80.0,-80.0,0.0
END
```

The next routine in Chapter 5, `ratval`, does not have its own demonstration program. Instead, it is used in the demonstration programs for `pade` and `ratlsq` at the end of this chapter.

Ridders' method for numerical differentiation is carried out by routine `dfridr`. In sample program `xdfridr` we test it on the simple function $f(x) = \tan x$, for which $f'(x) = \sec^2 x$. You might try input values like $x = 1.0$, $h = 0.1$, to demonstrate that h need not (and should not) be very small.

```
      PROGRAM xdfridr
C     driver for routine dfridr
      REAL dfridr,dx,err,func,h,x
      EXTERNAL func
1     write(*,*) 'input x,h '
      read(*,*,END=999) x,h
      dx=dfridr(func,x,h,err)
      write(*,*) 'DFRIDR=',dx,1./cos(x)**2,err
      goto 1
999   END

      REAL FUNCTION func(x)
      REAL x
      func=tan(x)
      return
      END
```

The next seven programs all deal with Chebyshev polynomials. `chebft` evaluates the coefficients for a Chebyshev polynomial approximation of a function on a specified interval and for a maximum degree N of polynomial. Demonstration program `xchebft` uses the function `func` $= x^2(x^2 - 2)\sin x$ on the interval $(-\pi/2, \pi/2)$ with the maximum degree of `NVAL=40`. Notice that `chebft` is called with this maximum degree specified, even though subsequent evaluations may truncate the Chebyshev series at much lower terms. After we choose the number `mval` of terms in the evaluation, the Chebyshev polynomial is evaluated term by term, for x values between -0.8π and 0.8π, and the result `f` is compared to the actual function value.

```
      PROGRAM xchebft
C     driver for routine chebft
      INTEGER NVAL
      REAL PIO2,EPS
      PARAMETER(NVAL=40, PIO2=1.5707963, EPS=1E-6)
      INTEGER i,j,mval
      REAL a,b,dum,f,func,t0,t1,term,x,y,c(NVAL)
      EXTERNAL func
      a=-PIO2
      b=PIO2
      call chebft(a,b,c,NVAL,func)
C     test result
```

```
10    write(*,*) 'How many terms in Chebyshev evaluation?'
      write(*,'(1x,a,i2,a)') 'Enter n between 6 and ',NVAL,
     *      '. Enter n=0 to END.'
      read(*,*) mval
      if ((mval.le.0).or.(mval.gt.NVAL)) goto 20
      write(*,'(1x,t10,a,t19,a,t28,a)') 'X','Actual','Chebyshev fit'
      do 12 i=-8,8,1
        x=i*PIO2/10.0
        y=(x-0.5*(b+a))/(0.5*(b-a))
C       evaluate Chebyshev polynomial without using routine CHEBEV
        t0=1.0
        t1=y
        f=c(2)*t1+c(1)*0.5
        do 11 j=3,mval
          dum=t1
          t1=2.0*y*t1-t0
          t0=dum
          term=c(j)*t1
          f=f+term
11      continue
        write(*,'(1x,3f12.6)') x,func(x),f
12    continue
      goto 10
20    END

      REAL FUNCTION func(x)
      REAL x
      func=(x**2)*(x**2-2.0)*sin(x)
      END
```

chebev is the Chebyshev polynomial evaluator and the next sample program xchebev uses it for the same problem just discussed. In fact, the program is identical except that it replaces the internal polynomial summation with chebev, which applies Clenshaw's recurrence to find the polynomial values.

```
      PROGRAM xchebev
C     driver for routine chebev
      INTEGER NVAL
      REAL PIO2
      PARAMETER(NVAL=40, PIO2=1.5707963)
      INTEGER i,mval
      REAL a,b,chebev,func,x,c(NVAL)
      EXTERNAL func
      a=-PIO2
      b=PIO2
      call chebft(a,b,c,NVAL,func)
C     test Chebyshev evaluation routine
10    write(*,*) 'How many terms in Chebyshev evaluation?'
      write(*,'(1x,a,i2,a)') 'Enter n between 6 and ',NVAL,
     *      '. Enter n=0 to END.'
      read(*,*) mval
      if ((mval.le.0).or.(mval.gt.NVAL)) goto 20
      write(*,'(1x,t10,a,t19,a,t28,a)') 'X','Actual','Chebyshev fit'
      do 11 i=-8,8,1
        x=i*PIO2/10.0
        write(*,'(1x,3f12.6)') x,func(x),chebev(a,b,c,mval,x)
11    continue
```

```
      goto 10
20    END


      REAL FUNCTION func(x)
      REAL x
      func=(x**2)*(x**2-2.0)*sin(x)
      END
```

By the same token, the tests for `chint` and `chder` needn't be much different. `chint` determines Chebyshev coefficients for the integral of the function, and `chder` for the derivative of the function, given the Chebyshev coefficients for the function itself (from `chebft`) and the interval (A, B) of evaluation. When applied to the function above, the true integral is

$$\texttt{fint} = 4x(x^2 - 7)\sin x - (x^4 - 14x^2 + 28)\cos x$$

and the true derivative is

$$\texttt{fder} = 4x(x^2 - 1)\sin x + x^2(x^2 - 2)\cos x$$

The code in sample programs `xchint` and `xchder` compares the true and Chebyshev-derived integral and derivative values for a range of x in the interval of evaluation. Since `chint` and `chder` return Chebyshev coefficients, and not the integral and derivative values themselves, calls to `chebev` are required for the comparison.

```
      PROGRAM xchint
C     driver for routine chint
      INTEGER NVAL
      REAL PIO2
      PARAMETER(NVAL=40, PIO2=1.5707963)
      INTEGER i,mval
      REAL a,b,chebev,fint,func,x,c(NVAL),cint(NVAL)
      EXTERNAL fint,func
      a=-PIO2
      b=PIO2
      call chebft(a,b,c,NVAL,func)
C     test integral
10    write(*,*) 'How many terms in Chebyshev evaluation?'
      write(*,'(1x,a,i2,a)') 'Enter n between 6 and ',NVAL,
     *     '. Enter n=0 to END.'
      read(*,*) mval
      if ((mval.le.0).or.(mval.gt.NVAL)) goto 20
      call chint(a,b,c,cint,mval)
      write(*,'(1x,t10,a,t19,a,t29,a)') 'X','Actual','Cheby. Integ.'
      do 11 i=-8,8,1
        x=i*PIO2/10.0
        write(*,'(1x,3f12.6)') x,fint(x)-fint(-PIO2),
     *         chebev(a,b,cint,mval,x)
11    continue
      goto 10
20    END


      REAL FUNCTION func(x)
      REAL x
      func=(x**2)*(x**2-2.0)*sin(x)
      END
```

```
      REAL FUNCTION fint(x)
C     integral of FUNC
      REAL x
      fint=4.0*x*((x**2)-7.0)*sin(x)-((x**4)-14.0*(x**2)+28.0)*cos(x)
      END


      PROGRAM xchder
C     driver for routine chder
      INTEGER NVAL
      REAL PIO2
      PARAMETER(NVAL=40, PIO2=1.5707963)
      INTEGER i,mval
      REAL a,b,chebev,fder,func,x,c(NVAL),cder(NVAL)
      EXTERNAL func,fder
      a=-PIO2
      b=PIO2
      call chebft(a,b,c,NVAL,func)
C     test derivative
10    write(*,*) 'How many terms in Chebyshev evaluation?'
      write(*,'(1x,a,i2,a)') 'Enter n between 6 and ',NVAL,
     *       '. Enter n=0 to END.'
      read(*,*) mval
      if ((mval.le.0).or.(mval.gt.NVAL)) goto 20
      call chder(a,b,c,cder,mval)
      write(*,'(1x,t10,a,t19,a,t28,a)') 'X','Actual','Cheby. Deriv.'
      do 11 i=-8,8,1
        x=i*PIO2/10.0
        write(*,'(1x,3f12.6)') x,fder(x),chebev(a,b,cder,mval,x)
11    continue
      goto 10
20    END


      REAL FUNCTION func(x)
      REAL x
      func=(x**2)*(x**2-2.0)*sin(x)
      END


      REAL FUNCTION fder(x)
C     derivative of FUNC
      REAL x
      fder=4.0*x*((x**2)-1.0)*sin(x)+(x**2)*(x**2-2.0)*cos(x)
      END
```

The next two programs of this chapter turn the coefficients of a Chebyshev approximation into those of a polynomial approximation in the variable

$$y = \frac{x - \frac{1}{2}(B+A)}{\frac{1}{2}(B-A)}$$

(routine chebpc), or of a polynomial approximation in x itself (routine chebpc followed by pcshft). These procedures are discouraged for reasons discussed in *Numerical Recipes*, but should they serve some special purpose for you, we have at least warned that you will be sacrificing accuracy, particularly for polynomials above order 7 or 8. Sample program xchebpc calls chebft and chebpc to find polynomial coefficients in y for a truncated series. For a set of x values between $-\pi$ and π it

calculates y and then the terms of the y-polynomial, which are summed in variable
poly. Finally, poly is compared to the true function value. (The function func is
the same used before.)

```
      PROGRAM xchebpc
C     driver for routine chebpc
      INTEGER NVAL
      REAL PIO2
      PARAMETER(NVAL=40, PIO2=1.5707963)
      INTEGER i,j,mval
      REAL a,b,func,poly,x,y,c(NVAL),d(NVAL)
      EXTERNAL func
      a=-PIO2
      b=PIO2
      call chebft(a,b,c,NVAL,func)
10    write(*,*) 'How many terms in Chebyshev evaluation?'
      write(*,'(1x,a,i2,a)') 'Enter n between 6 and ',NVAL,
     *      '. Enter n=0 to END.'
      read(*,*) mval
      if ((mval.le.0).or.(mval.gt.NVAL)) goto 20
      call chebpc(c,d,mval)
C     test polynomial
      write(*,'(1x,t10,a,t19,a,t29,a)') 'X','Actual','Polynomial'
      do 12 i=-8,8,1
        x=i*PIO2/10.0
        y=(x-(0.5*(b+a)))/(0.5*(b-a))
        poly=d(mval)
        do 11 j=mval-1,1,-1
          poly=poly*y+d(j)
11      continue
        write(*,'(1x,3f12.6)') x,func(x),poly
12    continue
      goto 10
20    END

      REAL FUNCTION func(x)
      REAL x
      func=(x**2)*(x**2-2.0)*sin(x)
      END
```

pcshft shifts the polynomial to be one in variable x. Sample program xpcshft
is like the previous program except that it follows the call to chebpc with a call to
pcshft.

```
      PROGRAM xpcshft
C     driver for routine pcshft
      INTEGER NVAL
      REAL PIO2
      PARAMETER(NVAL=40, PIO2=1.5707963)
      INTEGER i,j,mval
      REAL a,b,func,poly,x,c(NVAL),d(NVAL)
      EXTERNAL func
      a=-PIO2
      b=PIO2
      call chebft(a,b,c,NVAL,func)
10    write(*,*) 'How many terms in Chebyshev evaluation?'
      write(*,'(1x,a,i2,a)') 'Enter n between 6 and ',NVAL,
```

```
      *        '. Enter n=0 to END.'
      read(*,*) mval
      if ((mval.le.0).or.(mval.gt.NVAL)) goto 20
      call chebpc(c,d,mval)
      call pcshft(a,b,d,mval)
C     test shifted polynomial
      write(*,'(1x,t10,a,t19,a,t29,a)') 'X','Actual','Polynomial'
      do 12 i=-8,8,1
        x=i*PIO2/10.0
        poly=d(mval)
        do 11 j=mval-1,1,-1
          poly=poly*x+d(j)
11      continue
        write(*,'(1x,3f12.6)') x,func(x),poly
12    continue
      goto 10
20    END

      REAL FUNCTION func(x)
      REAL x
      func=(x**2)*(x**2-2.0)*sin(x)
      END
```

Program `xpccheb` shows how to use `pccheb` to economize a power series. The example chosen is the series on p. 294 of *Numerical Methods That Work*, by F.S. Acton, namely that of $\cos \pi y$ on the interval $[-1, 1]$. The coefficients of the series up to y^{17} are computed and converted to Chebyshev coefficients. These coefficients are truncated after the term in y^{13}, and the new power series coefficients are computed. The answer is checked at selected points in the interval.

```
      PROGRAM xpccheb
C     driver for routine pccheb
      INTEGER NCHECK,NFEW,NMANY,NMAX
      REAL PI
      PARAMETER(NCHECK=15,NFEW=13,NMANY=17,NMAX=100,PI=3.14159265)
      INTEGER i,j
      REAL a,b,fac,f,sum,sume,py,py2,c(NMAX),d(NMAX),e(NMAX),ee(NMAX)
C     put power series of cos(PI*y) into e
      fac=1.
      do 11 j=1,NMANY
        i=mod(j-1,4)
        if (i.eq.1.or.i.eq.3) then
          e(j)=0.
        else if (i.eq.0) then
          e(j)=1./fac
        else
          e(j)=-1./fac
        endif
        fac=fac*j
        ee(j)=e(j)
11    continue
      a=-PI
      b=PI
      call pcshft((-2.-b-a)/(b-a),(2.-b-a)/(b-a),e,NMANY)
C     i.e., inverse of PCSHFT(A,B,...) which we do below
      call pccheb(e,c,NMANY)
```

```
      write(*,*) 'Index, series, Chebyshev coefficients'
      do 12 j=1,NMANY,2
        write(*,'(i3,2e15.6)') j,e(j),c(j)
12    continue
      call chebpc(c,d,NFEW)
      call pcshft(a,b,d,NFEW)
      write(*,*) 'Index, new series, coefficient ratios'
      do 13 j=1,NFEW,2
        write(*,'(i3,2e15.6)') j,d(j),d(j)/(ee(j)+1.e-30)
13    continue
      write(*,'(7x,a)')
     *'Point tested, function value, error power series, error Cheb.'
      do 15 i=0,15
        py=(a+i*(b-a)/15.)
        py2=py*py
        sum=0.
        sume=0.
        fac=1.
        do 14 j=1,NFEW,2
          sum=sum+fac*d(j)
          sume=sume+fac*ee(j)
          fac=fac*py2
14      continue
        f=cos(py)
        write(*,'(1x,a,4e15.6)') 'check:',py,f,sume-f,sum-f
15    continue
      END
```

Test program `xpade` excercises `pade` by computing the Padé approximation to $\ln(1 + x)/x$. The Padé approximation is accurate for a much larger value of x than the power series with the same number of terms from which it is derived. Note that this test program and the one following both use `ratval`, so we do not supply a separate sample program for it.

```
      PROGRAM xpade
C     driver for routine pade
      INTEGER NMAX
      PARAMETER(NMAX=100)
      INTEGER j,k,n
      REAL resid
      DOUBLE PRECISION b,d,fac,fn,ratval,x,c(NMAX),cc(NMAX)
1     write(*,*) 'Enter n for PADE routine:'
      read(*,*,END=999) n
      fac=1
      do 11 j=1,2*n+1
        c(j)=fac/dble(j)
        cc(j)=c(j)
        fac=-fac
11    continue
      call pade(c,n,resid)
      write(*,'(1x,a,1pd16.8)') 'Norm of residual vector=',resid
      write(*,*) 'point, func. value, pade series, power series'
      do 13 j=1,21
        x=(j-1)*0.25
        b=0.
        do 12 k=2*n+1,1,-1
          b=b*x+cc(k)
```

```
12      continue
        d=ratval(x,c,n,n)
        write(*,'(1p4d16.8)') x,fn(x),d,b
13      continue
        goto 1
999     END

        DOUBLE PRECISION FUNCTION fn(x)
        DOUBLE PRECISION x
        if (x.eq.0.) then
          fn=1.
        else
          fn=log(1.+x)/x
        endif
        return
        END
```

Sample program `xratlsq` finds a rational approximation to the function $\tan^{-1} x$ by the least squares method in `ratlsq`. You enter a desired interval and the orders of the numerator and the denominator. The printed result shows the discrepancy between the fit and the true function value.

```
        PROGRAM xratlsq
C       driver for routine ratlsq
        INTEGER NMAX
        PARAMETER(NMAX=100)
        INTEGER j,kk,mm
        DOUBLE PRECISION a,b,cof(NMAX),dev,eee,fit,fn,ratval,xs
        EXTERNAL fn
1       write(*,*) 'enter a,b,mm,kk'
        read(*,*,END=999) a,b,mm,kk
        call ratlsq(fn,a,b,mm,kk,cof,dev)
        do 11 j=1,mm+kk+1
          write(*,'(1x,a4,i3,a2,e27.15)') 'cof(',j,')=',cof(j)
11      continue
        write(*,*) 'maximum absolute deviation=',dev
        write(*,*) '    x          error         exact'
        write(*,*) '--------   ------------   ---------'
        do 12 j=1,50
          xs=a+(b-a)*(j-1.)/49.
          fit=ratval(xs,cof,mm,kk)
          eee=fn(xs)
          write(*,'(1x,f10.5,2e15.7)') xs,fit-eee,eee
12      continue
        goto 1
999     END

        FUNCTION fn(t)
        DOUBLE PRECISION fn,t
        fn=atan(t)
        return
        END
```

Chapter 6: Special Functions

This chapter on special functions provides illustrations of techniques developed in Chapter 5. At the same time, it offers routines for calculating many of the functions that arise frequently in analytical work, but which are not so common as to be included, for example, as a single keystroke on your pocket calculator. In terms of demonstration programs, they represent a simple collection. The test routines are all virtually identical, making reference to a single file of function values called FNCVAL.DAT *which is listed in the Appendix at the end of this chapter. In this file are accurate values for the individual functions for a variety of values for each argument. We have aimed to "stress" the routines a bit by throwing in some extreme values for the arguments.*

Many of the function values came from Abramowitz and Stegun's Handbook of Mathematical Functions. Some others, however, came from our library of dusty volumes from past masters. There is an implicit danger in a comparison test like this — namely, that our source has used the same algorithms as ours to construct the tables. In that case, we test only our mutual competence at computing, not the correctness of the result. Nevertheless, there is some assurance in knowing that the values we calculate are the ones that have been used and scrutinized for many years. Moreover, the expressions for the functions themselves can be worked out in certain special or limiting cases without computer aid, and in these instances the results have proven correct.

$$\star \quad \star \quad \star \quad \star$$

With few exceptions, the routines that follow work in this fashion:

1. Open file FNCVAL.DAT.

2. Find the appropriate data table according to its title.

3. Read the argument list for each table entry and pass it to the routine to be tested.

4. Print the arguments along with the expected and actual results.

For the routines in this list, therefore, we forego any further comment, but simply identify them by the special function that they evaluate. The exceptions are the test program for the utility subroutine beschb and the test of the hypergeometric function routine hypgeo.

Natural logarithm of the gamma function for positive arguments:

```
      PROGRAM xgammln
C     driver for routine gammln
      REAL PI
      PARAMETER(PI=3.1415926)
      INTEGER i,nval
      REAL actual,calc,gammln,x
      CHARACTER text*14
      open(7,file='FNCVAL.DAT',status='OLD')
10    read(7,'(a)') text
      if (text.ne.'Gamma Function') goto 10
      read(7,*) nval
      write(*,*) 'Log of gamma function:'
      write(*,'(1x,t11,a1,t24,a6,t40,a10)')
     *    'X','Actual','GAMMLN(X)'
      do 11 i=1,nval
        read(7,*) x,actual
        if (x.gt.0.0) then
          if (x.ge.1.0) then
            calc=gammln(x)
          else
            calc=gammln(x+1.0)-log(x)
          endif
          write(*,'(f12.2,2f18.6)') x,log(actual),calc
        endif
11    continue
      close(7)
      END
```

Factorial function $N!$:

```
      PROGRAM xfactrl
C     driver for routine factrl
      INTEGER i,n,nval
      REAL actual,factrl
      CHARACTER text*11
      open(7,file='FNCVAL.DAT',status='OLD')
10    read(7,'(a)') text
      if (text.ne.'N-factorial') goto 10
      read(7,*) nval
      write(*,*) text
      write(*,'(1x,t6,a1,t21,a6,t38,a9)')
     *    'N','Actual','FACTRL(N)'
      do 11 i=1,nval
        read(7,*) n,actual
        if (actual.lt.(1.0e10)) then
          write(*,'(i6,2f20.0)') n,actual,factrl(n)
        else
          write(*,'(i6,2e20.7)') n,actual,factrl(n)
        endif
11    continue
      close(7)
      END
```

Binomial coefficients:

```
      PROGRAM xbico
C     driver for routine bico
      INTEGER i,k,n,nval
```

```
      REAL binco,bico
      CHARACTER text*21
      open(7,file='FNCVAL.DAT',status='OLD')
10    read(7,'(a)') text
      if (text.ne.'Binomial Coefficients') goto 10
      read(7,*) nval
      write(*,*) text
      write(*,'(1x,t6,a1,t12,a1,t19,a6,t28,a9)')
     *      'N','K','Actual','BICO(N,K)'
      do 11 i=1,nval
        read(7,*) n,k,binco
        write(*,'(2i6,2f12.0)') n,k,binco,bico(n,k)
11    continue
      close(7)
      END
```

Natural logarithm of $N!$:

```
      PROGRAM xfactln
C     driver for routine factln
      INTEGER i,n,nval
      REAL factln,value
      CHARACTER text*11
      open(7,file='FNCVAL.DAT',status='OLD')
10    read(7,'(a)') text
      if (text.ne.'N-factorial') goto 10
      read(7,*) nval
      write(*,*) 'Log of N-factorial'
      write(*,'(1x,t6,a1,t18,a6,t34,a9)')
     *      'N','Actual','FACTLN(N)'
      do 11 i=1,nval
        read(7,*) n,value
        write(*,'(i6,2f18.6)') n,log(value),factln(n)
11    continue
      close(7)
      END
```

Beta function:

```
      PROGRAM xbeta
C     driver for routine beta
      INTEGER i,nval
      REAL beta,value,w,z
      CHARACTER text*13
      open(7,file='FNCVAL.DAT',status='OLD')
10    read(7,'(a)') text
      if (text.ne.'Beta Function') goto 10
      read(7,*) nval
      write(*,*) text
      write(*,'(1x,t5,a1,t11,a1,t24,a6,t43,a9)')
     *      'W','Z','Actual','BETA(W,Z)'
      do 11 i=1,nval
        read(7,*) w,z,value
        write(*,'(2f6.2,2e20.6)') w,z,value,beta(w,z)
11    continue
      close(7)
      END
```

Incomplete gamma function $P(a, x)$:

```
      PROGRAM xgammp
C     driver for routine gammp
      INTEGER i,nval
      REAL a,gammp,value,x
      CHARACTER text*25
      open(7,file='FNCVAL.DAT',status='OLD')
10    read(7,'(a)') text
      if (text.ne.'Incomplete Gamma Function') goto 10
      read(7,*) nval
      write(*,*) text
      write(*,'(1x,t5,a,t16,a,t25,a,t35,a)')
     *     'A','X','Actual','GAMMP(A,X)'
      do 11 i=1,nval
        read(7,*) a,x,value
        write(*,'(1x,f6.2,3f12.6)') a,x,value,gammp(a,x)
11    continue
      close(7)
      END
```

Incomplete gamma function $Q(a, x) = 1 - P(a, x)$:

```
      PROGRAM xgammq
C     driver for routine gammq
      INTEGER i,nval
      REAL a,gammq,value,x
      CHARACTER text*25
      open(7,file='FNCVAL.DAT',status='OLD')
10    read(7,'(a)') text
      if (text.ne.'Incomplete Gamma Function') goto 10
      read(7,*) nval
      write(*,*) text
      write(*,'(1x,t5,a,t16,a,t25,a,t35,a)')
     *     'A','X','Actual','GAMMQ(A,X)'
      do 11 i=1,nval
        read(7,*) a,x,value
        write(*,'(1x,f6.2,3f12.6)') a,x,1.0-value,gammq(a,x)
11    continue
      close(7)
      END
```

Incomplete gamma function $P(a, x)$ evaluated from series representation:

```
      PROGRAM xgser
C     driver for routine gser
      INTEGER i,nval
      REAL a,gammln,gamser,gln,value,x
      CHARACTER text*25
      open(7,file='FNCVAL.DAT',status='OLD')
10    read(7,'(a)') text
      if (text.ne.'Incomplete Gamma Function') goto 10
      read(7,*) nval
      write(*,*) text
      write(*,'(1x,t5,a,t16,a,t25,a,t36,a,t47,a,t62,a)')
     *     'A','X','Actual','GSER(A,X)','GAMMLN(A)','GLN'
      do 11 i=1,nval
        read(7,*) a,x,value
```

```
        call gser(gamser,a,x,gln)
        write(*,'(1x,f6.2,5f12.6)') a,x,value,gamser,
     *      gammln(a),gln
11      continue
        close(7)
        END
```

Incomplete gamma function $Q(a, x)$ evaluated by continued fraction representation:

```
        PROGRAM xgcf
C       driver for routine gcf
        INTEGER i,nval
        REAL a,gammcf,gammln,gln,value,x
        CHARACTER text*25
        open(7,file='FNCVAL.DAT',status='OLD')
10      read(7,'(a)') text
        if (text.ne.'Incomplete Gamma Function') goto 10
        read(7,*) nval
        write(*,*) text
        write(*,'(1x,t5,a,t16,a,t25,a,t36,a,t47,a,t62,a)')
     *      'A','X','Actual','GCF(A,X)','GAMMLN(A)','GLN'
        do 11 i=1,nval
          read(7,*) a,x,value
          if (x.ge.a+1.0) then
            call gcf(gammcf,a,x,gln)
            write(*,'(1x,f6.2,5f12.6)') a,x,1.0-value,
     *          gammcf,gammln(a),gln
          endif
11      continue
        close(7)
        END
```

Error function:

```
        PROGRAM xerf
C       driver for routine erf
        INTEGER i,nval
        REAL erf,value,x
        CHARACTER text*14
        open(7,file='FNCVAL.DAT',status='OLD')
10      read(7,'(a)') text
        if (text.ne.'Error Function') goto 10
        read(7,*) nval
        write(*,*) text
        write(*,'(1x,t5,a1,t12,a6,t24,a6)')
     *      'X','Actual','ERF(X)'
        do 11 i=1,nval
          read(7,*) x,value
          write(*,'(f6.2,2f12.7)') x,value,erf(x)
11      continue
        close(7)
        END
```

Complementary error function:

```
        PROGRAM xerfc
C       driver for routine erfc
        INTEGER i,nval
```

```
      REAL erfc,value,x
      CHARACTER text*14
      open(7,file='FNCVAL.DAT',status='OLD')
10    read(7,'(a)') text
      if (text.ne.'Error Function') goto 10
      read(7,*) nval
      write(*,*) 'Complementary error function'
      write(*,'(1x,t5,a1,t12,a6,t23,a7)')
     *      'X','Actual','ERFC(X)'
      do 11 i=1,nval
        read(7,*) x,value
        value=1.0-value
        write(*,'(f6.2,2f12.7)') x,value,erfc(x)
11    continue
      close(7)
      END
```

Complementary error function from a Chebyshev fit to a guessed functional form:

```
      PROGRAM xerfcc
C     driver for routine erfcc
      INTEGER i,nval
      REAL erfcc,value,x
      CHARACTER text*14
      open(7,file='FNCVAL.DAT',status='OLD')
10    read(7,'(a)') text
      if (text.ne.'Error Function') goto 10
      read(7,*) nval
      write(*,*) 'Complementary error function'
      write(*,'(1x,t5,a1,t12,a6,t23,a8)')
     *      'X','Actual','ERFCC(X)'
      do 11 i=1,nval
        read(7,*) x,value
        value=1.0-value
        write(*,'(f6.2,2f12.7)') x,value,erfcc(x)
11    continue
      close(7)
      END
```

Exponential integral $E_n(x)$:

```
      PROGRAM xexpint
C     driver for routine expint
      INTEGER i,n,nval
      REAL expint,value,x
      CHARACTER text*23
      open(7,file='FNCVAL.DAT',status='OLD')
10    read(7,'(a)') text
      if (text.ne.'Exponential Integral En') goto 10
      read(7,*) nval
      write(*,*) text
      write(*,'(1x,t5,a,t12,a,t20,a,t33,a)')
     *      'N','X','Actual','EXPINT(N,X)'
      do 11 i=1,nval
        read(7,*) n,x,value
        write(*,'(1x,i4,f8.2,2e15.6)') n,x,value,expint(n,x)
11    continue
      close(7)
```

```
      END
```

Exponential integral Ei(x):

```
      PROGRAM xei
C     driver for routine ei
      INTEGER i,nval
      REAL ei,value,x
      CHARACTER text*23
      open(7,file='FNCVAL.DAT',status='OLD')
10    read(7,'(a)') text
      if (text.ne.'Exponential Integral Ei') goto 10
      read(7,*) nval
      write(*,*) text
      write(*,'(1x,t5,a,t13,a,t28,a)')
     *     'X','Actual','EI(X)'
      do 11 i=1,nval
         read(7,*) x,value
         write(*,'(f6.2,2e16.6)') x,value,ei(x)
11    continue
      close(7)
      END
```

Incomplete Beta function:

```
      PROGRAM xbetai
C     driver for routine betai,betacf
      INTEGER i,nval
      REAL a,b,betai,value,x
      CHARACTER text*24
      open(7,file='FNCVAL.DAT',status='OLD')
10    read(7,'(a)') text
      if (text.ne.'Incomplete Beta Function') goto 10
      read(7,*) nval
      write(*,*) text
      write(*,'(1x,t5,a,t15,a,t27,a,t36,a,t47,a)')
     *     'A','B','X','Actual','BETAI(X)'
      do 11 i=1,nval
         read(7,*) a,b,x,value
         write(*,'(f6.2,4f12.6)') a,b,x,value,betai(a,b,x)
11    continue
      close(7)
      END
```

Bessel function J_0:

```
      PROGRAM xbessj0
C     driver for routine bessj0
      INTEGER i,nval
      REAL bessj0,value,x
      CHARACTER text*18
      open(7,file='FNCVAL.DAT',status='OLD')
10    read(7,'(a)') text
      if (text.ne.'Bessel Function J0') goto 10
      read(7,*) nval
      write(*,*) text
      write(*,'(1x,t5,a1,t12,a6,t22,a9)')
     *     'X','Actual','BESSJ0(X)'
```

```
      do 11 i=1,nval
        read(7,*) x,value
        write(*,'(f6.2,2f12.7)') x,value,bessj0(x)
11    continue
      close(7)
      END
```

Bessel function Y_0:

```
      PROGRAM xbessy0
C     driver for routine bessy0
      INTEGER i,nval
      REAL bessy0,value,x
      CHARACTER text*18
      open(7,file='FNCVAL.DAT',status='OLD')
10    read(7,'(a)') text
      if (text.ne.'Bessel Function Y0') goto 10
      read(7,*) nval
      write(*,*) text
      write(*,'(1x,t5,a1,t12,a6,t22,a9)')
     *     'X','Actual','BESSY0(X)'
      do 11 i=1,nval
        read(7,*) x,value
        write(*,'(f6.2,2f12.7)') x,value,bessy0(x)
11    continue
      close(7)
      END
```

Bessel function J_1:

```
      PROGRAM xbessj1
C     driver for routine bessj1
      INTEGER i,nval
      REAL bessj1,value,x
      CHARACTER text*18
      open(7,file='FNCVAL.DAT',status='OLD')
10    read(7,'(a)') text
      if (text.ne.'Bessel Function J1') goto 10
      read(7,*) nval
      write(*,*) text
      write(*,'(1x,t5,a1,t12,a6,t22,a9)')
     *     'X','Actual','BESSJ1(X)'
      do 11 i=1,nval
        read(7,*) x,value
        write(*,'(f6.2,2f12.7)') x,value,bessj1(x)
11    continue
      close(7)
      END
```

Bessel function Y_1:

```
      PROGRAM xbessy1
C     driver for routine bessy1
      INTEGER i,nval
      REAL bessy1,value,x
      CHARACTER text*18
      open(7,file='FNCVAL.DAT',status='OLD')
10    read(7,'(a)') text
```

```
      if (text.ne.'Bessel Function Y1') goto 10
      read(7,*) nval
      write(*,*) text
      write(*,'(1x,t5,a1,t12,a6,t22,a9)')
     *    'X','Actual','BESSY1(X)'
      do 11 i=1,nval
        read(7,*) x,value
        write(*,'(f6.2,2f12.7)') x,value,bessy1(x)
11    continue
      close(7)
      END
```

Bessel function Y_n for $n > 1$:

```
      PROGRAM xbessy
C     driver for routine bessy
      INTEGER i,n,nval
      REAL bessy,value,x
      CHARACTER text*18
      open(7,file='FNCVAL.DAT',status='OLD')
10    read(7,'(a)') text
      if (text.ne.'Bessel Function Yn') goto 10
      read(7,*) nval
      write(*,*) text
      write(*,'(1x,t5,a,t12,a,t20,a,t33,a)')
     *    'N','X','Actual','BESSY(N,X)'
      do 11 i=1,nval
        read(7,*) n,x,value
        write(*,'(1x,i4,f8.2,2e15.6)') n,x,value,bessy(n,x)
11    continue
      close(7)
      END
```

Bessel function J_n for $n > 1$:

```
      PROGRAM xbessj
C     driver for routine bessj
      INTEGER i,n,nval
      REAL bessj,value,x
      CHARACTER text*18
      open(7,file='FNCVAL.DAT',status='OLD')
10    read(7,'(a)') text
      if (text.ne.'Bessel Function Jn') goto 10
      read(7,*) nval
      write(*,*) text
      write(*,'(1x,t5,a,t12,a,t20,a,t33,a)')
     *    'N','X','Actual','BESSJ(N,X)'
      do 11 i=1,nval
        read(7,*) n,x,value
        write(*,'(1x,i4,f8.2,2e15.6)') n,x,value,bessj(n,x)
11    continue
      close(7)
      END
```

Bessel function I_0:

```
      PROGRAM xbessi0
C     driver for routine bessi0
      INTEGER i,nval
      REAL bessi0,value,x
      CHARACTER text*27
      open(7,file='FNCVAL.DAT',status='OLD')
10    read(7,'(a)') text
      if (text.ne.'Modified Bessel Function I0') goto 10
      read(7,*) nval
      write(*,*) text
      write(*,'(1x,t5,a,t13,a,t28,a)')
     *      'X','Actual','BESSI0(X)'
      do 11 i=1,nval
        read(7,*) x,value
        write(*,'(f6.2,2e16.7)') x,value,bessi0(x)
11    continue
      close(7)
      END
```

Bessel function K_0:

```
      PROGRAM xbessk0
C     driver for routine bessk0
      INTEGER i,nval
      REAL bessk0,value,x
      CHARACTER text*27
      open(7,file='FNCVAL.DAT',status='OLD')
10    read(7,'(a)') text
      if (text.ne.'Modified Bessel Function K0') goto 10
      read(7,*) nval
      write(*,*) text
      write(*,'(1x,t5,a,t13,a,t28,a)')
     *      'X','Actual','BESSK0(X)'
      do 11 i=1,nval
        read(7,*) x,value
        write(*,'(f6.2,2e16.7)') x,value,bessk0(x)
11    continue
      close(7)
      END
```

Bessel function I_1:

```
      PROGRAM xbessi1
C     driver for routine bessi1
      INTEGER i,nval
      REAL bessi1,value,x
      CHARACTER text*27
      open(7,file='FNCVAL.DAT',status='OLD')
10    read(7,'(a)') text
      if (text.ne.'Modified Bessel Function I1') goto 10
      read(7,*) nval
      write(*,*) text
      write(*,'(1x,t5,a,t13,a,t28,a)')
     *      'X','Actual','BESSI1(X)'
      do 11 i=1,nval
        read(7,*) x,value
        write(*,'(f6.2,2e16.7)') x,value,bessi1(x)
11    continue
```

```
      close(7)
      END
```

Bessel function K_1:

```
      PROGRAM xbessk1
C     driver for routine bessk1
      INTEGER i,nval
      REAL bessk1,value,x
      CHARACTER text*27
      open(7,file='FNCVAL.DAT',status='OLD')
10    read(7,'(a)') text
      if (text.ne.'Modified Bessel Function K1') goto 10
      read(7,*) nval
      write(*,*) text
      write(*,'(1x,t5,a,t13,a,t28,a)')
     *     'X','Actual','BESSK1(X)'
      do 11 i=1,nval
        read(7,*) x,value
        write(*,'(f6.2,2e16.7)') x,value,bessk1(x)
11    continue
      close(7)
      END
```

Bessel function K_n for $n > 1$:

```
      PROGRAM xbessk
C     driver for routine bessk
      INTEGER i,n,nval
      REAL bessk,value,x
      CHARACTER text*27
      open(7,file='FNCVAL.DAT',status='OLD')
10    read(7,'(a)') text
      if (text.ne.'Modified Bessel Function Kn') goto 10
      read(7,*) nval
      write(*,*) text
      write(*,'(1x,t5,a,t12,a,t20,a,t35,a)')
     *     'N','X','Actual','BESSK(N,X)'
      do 11 i=1,nval
        read(7,*) n,x,value
        write(*,'(1x,i4,f8.2,2e16.7)') n,x,value,bessk(n,x)
11    continue
      close(7)
      END
```

Bessel function I_n for $n > 1$:

```
      PROGRAM xbessi
C     driver for routine bessi
      INTEGER i,n,nval
      REAL bessi,value,x
      CHARACTER text*27
      open(7,file='FNCVAL.DAT',status='OLD')
10    read(7,'(a)') text
      if (text.ne.'Modified Bessel Function In') goto 10
      read(7,*) nval
      write(*,*) text
      write(*,'(1x,t5,a,t12,a,t20,a,t34,a)')
```

```
      *      'N','X','Actual','BESSI(N,X)'
       do 11 i=1,nval
         read(7,*) n,x,value
         write(*,'(1x,i4,f8.2,2e16.7)') n,x,value,bessi(n,x)
11     continue
       close(7)
       END
```

Bessel functions of fractional order, J_ν and Y_ν:

```
       PROGRAM xbessjy
C      driver for routine bessjy
       INTEGER i,nval
       REAL rj,ry,rjp,ryp,x,xnu,xrj,xry,xrjp,xryp
       CHARACTER text*25
       open(7,file='FNCVAL.DAT',status='OLD')
10     read(7,'(a)') text
       if (text.ne.'Ordinary Bessel Functions') goto 10
       read(7,*) nval
       write(*,*) text
       write(*,'(1x,t3,a3,t8,a1)') 'XNU','X'
       write(*,'(1x,t5,a2,t21,a2,t37,a3,t53,a3)') 'RJ','RY','RJP','RYP'
       write(*,'(1x,t5,a3,t21,a3,t37,a4,t53,a4)')
      *      'XRJ','XRY','XRJP','XRYP'
       do 11 i=1,nval
         read(7,*) xnu,x,rj,ry,rjp,ryp
         call bessjy(x,xnu,xrj,xry,xrjp,xryp)
         write(*,'(2f6.2/1p4e16.6/1p4e16.6)')
      *        xnu,x,rj,ry,rjp,ryp,xrj,xry,xrjp,xryp
11     continue
       close(7)
       END
```

The routine `bessjy` calls the utility procedure `beschb` to calculate the following quantities:

$$\Gamma_+ \equiv \frac{1}{\Gamma(1+x)}, \qquad \Gamma_- \equiv \frac{1}{\Gamma(1-x)}$$
$$\Gamma_1 \equiv \frac{1}{2x}(\Gamma_- - \Gamma_+), \qquad \Gamma_2 \equiv \frac{1}{2}(\Gamma_- + \Gamma_+)$$

Test program `xbeschb` compares the result of calling `beschb` with the results obtained by calling `gammln`.

```
       PROGRAM xbeschb
C      driver for routine beschb
       DOUBLE PRECISION gam1,gam2,gamp1,gammi,x,xgam1,xgam2,xgamp1,xgammi
       REAL gammln
1      write(*,*) 'Enter X'
       read(*,*,END=999) x
       call beschb(x,xgam1,xgam2,xgamp1,xgammi)
       write(*,'(1x,t3,a1)') 'X'
       write(*,'(1x,t5,a4,t21,a4,t37,a5,t53,a5)')
      *      'GAM1','GAM2','GAMP1','GAMMI'
       write(*,'(1x,t5,a4,t21,a4,t37,a5,t53,a5)')
      *      'XGAM1','XGAM2','XGAMP1','XGAMMI'
       gamp1=1/exp(gammln(1+sngl(x)))
       gammi=1/exp(gammln(1-sngl(x)))
```

```
        gam1=(gammi-gamp1)/(2*x)
        gam2=(gammi+gamp1)/2
        write(*,'(f6.2/1p4e16.6/1p4e16.6)')
     *      x,gam1,gam2,gamp1,gammi,xgam1,xgam2,xgamp1,xgammi
        go to 1
999     END
```

Modified Bessel functions of fractional order, I_ν and K_ν:

```
        PROGRAM xbessik
C       driver for routine bessik
        INTEGER i,nval
        REAL ri,rk,rip,rkp,x,xnu,xri,xrk,xrip,xrkp
        CHARACTER text*25
        open(7,file='FNCVAL.DAT',status='OLD')
10      read(7,'(a)') text
        if (text.ne.'Modified Bessel Functions') goto 10
        read(7,*) nval
        write(*,*) text
        write(*,'(1x,t3,a3,t8,a1)') 'XNU','X'
        write(*,'(1x,t5,a2,t21,a2,t37,a3,t53,a3)') 'RI','RK','RIP','RKP'
        write(*,'(1x,t5,a3,t21,a3,t37,a4,t53,a4)')
     *      'XRI','XRK','XRIP','XRKP'
        do 11 i=1,nval
          read(7,*) xnu,x,ri,rk,rip,rkp
          call bessik(x,xnu,xri,xrk,xrip,xrkp)
          write(*,'(2f6.2/1p4e16.6/1p4e16.6)')
     *          xnu,x,ri,rk,rip,rkp,xri,xrk,xrip,xrkp
11      continue
        close(7)
        END
```

Airy functions:

```
        PROGRAM xairy
C       driver for routine airy
        INTEGER i,nval
        REAL ai,bi,aip,bip,x,xai,xbi,xaip,xbip
        CHARACTER text*14
        open(7,file='FNCVAL.DAT',status='OLD')
10      read(7,'(a)') text
        if (text.ne.'Airy Functions') goto 10
        read(7,*) nval
        write(*,*) text
        write(*,'(1x,t3,a1)') 'X'
        write(*,'(1x,t5,a2,t21,a2,t37,a3,t53,a3)') 'AI','BI','AIP','BIP'
        write(*,'(1x,t5,a3,t21,a3,t37,a4,t53,a4)')
     *    'XAI','XBI','XAIP','XBIP'
        do 11 i=1,nval
          read(7,*) x,ai,bi,aip,bip
          call airy(x,xai,xbi,xaip,xbip)
          write(*,'(f6.2/1p4e16.6/1p4e16.6)')
     *          x,ai,bi,aip,bip,xai,xbi,xaip,xbip
11      continue
        close(7)
        END
```

Spherical Bessel functions:

```
      PROGRAM xsphbes
C     driver for routine sphbes
      INTEGER i,n,nval
      REAL sj,sy,sjp,syp,x,xsj,xsy,xsjp,xsyp
      CHARACTER text*26
      open(7,file='FNCVAL.DAT',status='OLD')
10    read(7,'(a)') text
      if (text.ne.'Spherical Bessel Functions') goto 10
      read(7,*) nval
      write(*,*) text
      write(*,'(1x,t3,a1,t6,a1)') 'N','X'
      write(*,'(1x,t5,a2,t21,a2,t37,a3,t53,a3)') 'SJ','SY','SJP','SYP'
      write(*,'(1x,t5,a3,t21,a3,t37,a4,t53,a4)')
     *    'XSJ','XSY','XSJP','XSYP'
      do 11 i=1,nval
        read(7,*) n,x,sj,sy,sjp,syp
        call sphbes(n,x,xsj,xsy,xsjp,xsyp)
        write(*,'(i3,f6.2/1p4e16.6/1p4e16.6)')
     *      n,x,sj,sy,sjp,syp,xsj,xsy,xsjp,xsyp
11    continue
      close(7)
      END
```

Legendre polynomials:

```
      PROGRAM xplgndr
C     driver for routine plgndr
      INTEGER i,j,m,n,nval
      REAL fac,plgndr,value,x
      CHARACTER text*27
      open(7,file='FNCVAL.DAT',status='OLD')
10    read(7,'(a)') text
      if (text.ne.'Legendre Polynomials') go to 10
      read(7,*) nval
      write(*,*) text
      write(*,'(1x,t5,a,t9,a,t20,a,t35,a,t49,a)')
     *    'N','M','X','Actual','PLGNDR(N,M,X)'
      do 12 i=1,nval
        read(7,*) n,m,x,value
        fac=1.0
        if (m.gt.0) then
          do 11 j=n-m+1,n+m
            fac=fac*j
11        continue
        endif
        fac=2.0*fac/(2.0*n+1.0)
        value=value*sqrt(fac)
        write(*,'(1x,2i4,3e17.6)') n,m,x,value,plgndr(n,m,x)
12    continue
      close(7)
      END
```

Fresnel integrals:

```
      PROGRAM xfrenel
C     driver for routine frenel
      INTEGER i,nval
      REAL c,s,x,xc,xs
      CHARACTER text*17
      open(7,file='FNCVAL.DAT',status='OLD')
10    read(7,'(a)') text
      if (text.ne.'Fresnel Integrals') goto 10
      read(7,*) nval
      write(*,*) text
      write(*,'(1x,t5,a1,t12,a6,t28,a4,t42,a6,t58,a4)')
     *    'X','Actual','S(X)','Actual','C(X)'
      do 11 i=1,nval
        read(7,*) x,xs,xc
        call frenel(x,s,c)
        write(*,'(f6.2,4e15.6)') x,xs,s,xc,c
11    continue
      close(7)
      END
```

Cosine and sine integrals:

```
      PROGRAM xcisi
C     driver for routine cisi
      INTEGER i,nval
      REAL ci,si,x,xci,xsi
      CHARACTER text*25
      open(7,file='FNCVAL.DAT',status='OLD')
10    read(7,'(a)') text
      if (text.ne.'Cosine and Sine Integrals') goto 10
      read(7,*) nval
      write(*,*) text
      write(*,'(1x,t5,a1,t13,a6,t25,a5,t37,a6,t48,a5)')
     *    'X','Actual','CI(X)','Actual','SI(X)'
      do 11 i=1,nval
        read(7,*) x,xci,xsi
        call cisi(x,ci,si)
        write(*,'(f6.2,4f12.6)') x,xci,ci,xsi,si
11    continue
      close(7)
      END
```

Dawson's integral:

```
      PROGRAM xdawson
C     driver for routine dawson
      INTEGER i,nval
      REAL dawson,value,x
      CHARACTER text*15
      open(7,file='FNCVAL.DAT',status='OLD')
10    read(7,'(a)') text
      if (text.ne.'Dawson integral') goto 10
      read(7,*) nval
      write(*,*) text
      write(*,'(1x,t5,a,t13,a,t28,a)')
     *    'X','Actual','Dawson(X)'
      do 11 i=1,nval
        read(7,*) x,value
```

```
      write(*,'(f6.2,2e16.6)') x,value,dawson(x)
11    continue
      close(7)
      END
```

Carlson's elliptic integral of the first kind:

```
      PROGRAM xrf
C     driver for routine rf
      INTEGER i,nval
      REAL rf,value,x,y,z
      CHARACTER text*31
      open(7,file='FNCVAL.DAT',status='OLD')
10    read(7,'(a)') text
      if (text.ne.'Elliptic Integral First Kind RF') goto 10
      read(7,*) nval
      write(*,*) text
      write(*,'(1x,t7,a1,t15,a1,t23,a1,t36,a6,t54,a9)')
     *      'X','Y','Z','Actual','RF(X,Y,Z)'
      do 11 i=1,nval
        read(7,*) x,y,z,value
        write(*,'(3f8.2,2e20.6)') x,y,z,value,rf(x,y,z)
11    continue
      close(7)
      END
```

Carlson's elliptic integral of the second kind:

```
      PROGRAM xrd
C     driver for routine rd
      INTEGER i,nval
      REAL rd,value,x,y,z
      CHARACTER text*32
      open(7,file='FNCVAL.DAT',status='OLD')
10    read(7,'(a)') text
      if (text.ne.'Elliptic Integral Second Kind RD') goto 10
      read(7,*) nval
      write(*,*) text
      write(*,'(1x,t7,a1,t15,a1,t23,a1,t36,a6,t54,a9)')
     *      'X','Y','Z','Actual','RD(X,Y,Z)'
      do 11 i=1,nval
        read(7,*) x,y,z,value
        write(*,'(3f8.2,2e20.6)') x,y,z,value,rd(x,y,z)
11    continue
      close(7)
      END
```

Carlson's elliptic integral of the third kind:

```
      PROGRAM xrj
C     driver for routine rj
      INTEGER i,nval
      REAL p,rj,value,x,y,z
      CHARACTER text*31
      open(7,file='FNCVAL.DAT',status='OLD')
10    read(7,'(a)') text
      if (text.ne.'Elliptic Integral Third Kind RJ') goto 10
      read(7,*) nval
```

```
      write(*,*) text
      write(*,'(1x,t7,a1,t15,a1,t23,a1,t31,a1,t44,a6,t61,a11)')
     *     'X','Y','Z','P',' ','Actual','RJ(X,Y,Z,P)'
      do 11 i=1,nval
        read(7,*) x,y,z,p,value
        write(*,'(4f8.2,2e20.6)') x,y,z,p,value,rj(x,y,z,p)
11    continue
      close(7)
      END
```

Carlson's degenerate elliptic integral:

```
      PROGRAM xrc
C     driver for routine rc
      INTEGER i,nval
      REAL rc,value,x,y
      CHARACTER text*31
      open(7,file='FNCVAL.DAT',status='OLD')
10    read(7,'(a)') text
      if (text.ne.'Elliptic Integral Degenerate RC') goto 10
      read(7,*) nval
      write(*,*) text
      write(*,'(1x,t7,a1,t15,a1,t28,a6,t47,a7)')
     *     'X','Y','Actual','RC(X,Y)'
      do 11 i=1,nval
        read(7,*) x,y,value
        write(*,'(2f8.2,2e20.6)') x,y,value,rc(x,y)
11    continue
      close(7)
      END
```

Legendre elliptic integral of the first kind:

```
      PROGRAM xellf
C     driver for routine ellf
      REAL FAC
      PARAMETER (FAC=3.1415926535/180.)
      INTEGER i,nval
      REAL ak,alpha,ellf,phi,value
      CHARACTER text*37
      open(7,file='FNCVAL.DAT',status='OLD')
10    read(7,'(a)') text
      if (text.ne.'Legendre Elliptic Integral First Kind') goto 10
      read(7,*) nval
      write(*,*) text
      write(*,'(1x,t4,a3,t9,a10,t23,a6,t41,a12)')
     *     'PHI','SIN(ALPHA)','Actual','ELLF(PHI,AK)'
      do 11 i=1,nval
        read(7,*) phi,alpha,value
        alpha=alpha*FAC
        ak=sin(alpha)
        phi=phi*FAC
        write(*,'(2f6.2,2e20.6)') phi,ak,value,ellf(phi,ak)
11    continue
      close(7)
      END
```

Legendre elliptic integral of the second kind:

```
      PROGRAM xelle
C     driver for routine elle
      REAL FAC
      PARAMETER (FAC=3.1415926535/180.)
      INTEGER i,nval
      REAL ak,alpha,elle,phi,value
      CHARACTER text*38
      open(7,file='FNCVAL.DAT',status='OLD')
10    read(7,'(a)') text
      if (text.ne.'Legendre Elliptic Integral Second Kind') goto 10
      read(7,*) nval
      write(*,*) text
      write(*,'(1x,t4,a3,t9,a10,t23,a6,t41,a12)')
     *    'PHI','SIN(ALPHA)','Actual','ELLE(PHI,AK)'
      do 11 i=1,nval
        read(7,*) phi,alpha,value
        alpha=alpha*FAC
        ak=sin(alpha)
        phi=phi*FAC
        write(*,'(2f6.2,2e20.6)') phi,ak,value,elle(phi,ak)
11    continue
      close(7)
      END
```

Legendre elliptic integral of the third kind:

```
      PROGRAM xellpi
C     driver for routine ellpi
      REAL FAC
      PARAMETER (FAC=3.1415926535/180.)
      INTEGER i,nval
      REAL ak,alpha,ellpi,en,phi,value
      CHARACTER text*37
      open(7,file='FNCVAL.DAT',status='OLD')
10    read(7,'(a)') text
      if (text.ne.'Legendre Elliptic Integral Third Kind') goto 10
      read(7,*) nval
      write(*,*) text
      write(*,'(1x,t4,a3,t10,a3,t15,a10,t29,a6,t47,a16)')
     *    'PHI','-EN','SIN(ALPHA)','Actual','ELLPI(PHI,EN,AK)'
      do 11 i=1,nval
        read(7,*) phi,en,alpha,value
        alpha=alpha*FAC
        ak=sin(alpha)
        en=-en
        phi=phi*FAC
        write(*,'(3f6.2,2e20.6)') phi,en,ak,value,ellpi(phi,en,ak)
11    continue
      close(7)
      END
```

Routine sncndn returns Jacobian elliptic functions. The file FNCVAL.DAT contains information only about function sn. However, the values of cn and dn satisfy the relationships

$$\mathrm{sn}^2 + \mathrm{cn}^2 = 1, \qquad k^2\mathrm{sn}^2 + \mathrm{dn}^2 = 1.$$

The program `xsncndn` works exactly as the others in terms of testing `sn`, but for verifying `cn` and `dn` it lists the values `result1` and `result2` of the left sides of the two equations above. Each of them should have the value 1.0 for all choices of arguments.

```
      PROGRAM xsncndn
C     driver for routine sncndn
      INTEGER i,nval
      REAL cn,dn,em,emmc,resul1,resul2,sn,uu,value
      CHARACTER text*26
      open(7,file='FNCVAL.DAT',status='OLD')
10    read(7,'(a)') text
      if (text.ne.'Jacobian Elliptic Function') go to 10
      read(7,*) nval
      write(*,*) text
      write(*,'(1x,t4,a,t13,a,t21,a,t38,a,t49,a,t60,a)')
     *      'Mc','U','Actual','SN','SN^2+CN^2',
     *      '(Mc)*(SN^2)+DN^2'
      do 11 i=1,nval
        read(7,*) em,uu,value
        emmc=1.0-em
        call sncndn(uu,emmc,sn,cn,dn)
        resul1=sn*sn+cn*cn
        resul2=em*sn*sn+dn*dn
        write(*,'(1x,f5.2,f8.2,2e15.5,f12.5,f14.5)')
     *        emmc,uu,value,sn,resul1,resul2
11    continue
      close(7)
      END
```

The test program `xhypgeo` tests out the suite `hypgeo`, `hypdrv` and `hypser` by calculating a hypergeometric function that can be expressed in closed form:

$$_2F_1\left(\tfrac{1}{2},1,\tfrac{3}{2},z^2\right) = \frac{1}{2z}\log\frac{1+z}{1-z}$$

$$_2F_1\left(\tfrac{1}{2},1,\tfrac{3}{2},-z^2\right) = \frac{1}{2iz}\log\frac{1+iz}{1-iz}$$

The program prompts you to enter various values of x and y and compares the results for $z = x + iy$.

```
      PROGRAM xhypgeo
C     driver for routine hypgeo
      COMPLEX a,b,c,z,zi,hypgeo,q1,q2,q3,q4
      REAL x,y
      a=0.5
      b=1.
      c=1.5
1     write(*,*) 'INPUT X,Y OF COMPLEX ARGUMENT:'
      read(*,*,END=999) x,y
      z=cmplx(x,y)
      q1=hypgeo(a,b,c,z*z)
      q2=0.5*log((1.+z)/(1.-z))/z
      q3=hypgeo(a,b,c,-z*z)
      zi=cmplx(-y,x)
      q4=0.5*log((1.+zi)/(1.-zi))/zi
      write(*,*) '2F1(0.5,1.0,1.5;Z**2) =',q1
      write(*,*) 'check using log form   ',q2
```

```
        write(*,*) '2F1(0.5,1.0,1.5;-Z**2)=',q3
        write(*,*) 'check using log form   ',q4
        goto 1
999     END
```

Appendix

File FNCVAL.DAT:

```
Values of Special Functions in format x,F(x) or x,y,F(x,y)
Dawson integral
4 Values
0.04  0.0399573606
0.16  0.1572970920
1.6   0.3999398943
10.0  0.0502538471
Ordinary Bessel Functions       nu,x,jnu(x),ynu(x),jnup(x),ynup(x)
20 Values
2.0   1.0    1.149034849E-01 -1.650682607E+00  2.102436159E-01  2.520152392E+00
2.0   2.0    3.528340286E-01 -6.174081042E-01  2.238907791E-01  5.103756726E-01
2.0   5.0    4.656511628E-02  3.676628826E-01 -3.462051841E-01  7.979903490E-04
2.0   10.0   2.546303137E-01 -5.868082442E-03 -7.453316568E-03  2.501890407E-01
2.0   50.0  -5.971280079E-02  9.579316873E-02 -9.512331609E-02 -6.062739531E-02
5.0   1.0    2.497577302E-04 -2.604058666E+02  1.227850313E-03  1.268750910E+03
5.0   2.0    7.039629756E-03 -9.935989128E+00  1.639664542E-02  2.207402959E+01
5.0   5.0    2.611405461E-01 -4.536948225E-01  1.300918143E-01  2.615525351E-01
5.0   10.0  -2.340615282E-01  1.354030477E-01 -1.025719220E-01 -2.126510357E-01
5.0   50.0  -8.140024770E-02 -7.854841391E-02  7.898100205E-02 -8.020323269E-02
10.0  1.0    2.630615124E-10 -1.216180143E+08  2.618635056E-09  1.209399938E+09
10.0  2.0    2.515386283E-07 -1.291845422E+05  1.234650294E-06  6.313628817E+05
10.0  5.0    1.467802647E-03 -2.512911010E+01  2.584677845E-03  4.249433700E+01
10.0  10.0   2.074861066E-01 -3.598141522E-01  8.436957863E-02  1.605148864E-01
10.0  50.0  -1.138478491E-01  5.723897182E-03 -4.422891214E-03 -1.116145748E-01
20.0  1.0    3.873503009E-25 -4.113970315E+22  7.737778395E-24  8.217106466E+23
20.0  2.0    3.918972805E-19 -4.081651389E+16  3.900270468E-18  4.060105807E+17
20.0  5.0    2.770330052E-11 -5.933965297E+08  1.074693821E-10  2.294022549E+09
20.0  10.0   1.151336925E-05 -1.597483848E+03  2.011953903E-05  2.737803151E+03
20.0  50.0  -1.167043528E-01  1.644263395E-02 -1.368329398E-02 -1.071717186E-01
Modified Bessel Functions       nu,x,inu(x),knu(x),inup(x),knup(x)
28 Values
2.0   0.2    5.016687514E-03  4.951242929E+01  5.033395889E-02 -4.999002654E+02
2.0   1.0    1.357476698E-01  1.624838899E+00  2.936637645E-01 -3.851585027E+00
2.0   2.0    6.889484477E-01  2.537597546E-01  9.016884069E-01 -3.936256364E-01
2.0   2.5    1.276466148E+00  1.214602063E-01  1.495543327E+00 -1.710589814E-01
2.0   3.0    2.245212441E+00  6.151045847E-02  2.456561923E+00 -8.116340344E-02
2.0   5.0    1.750561497E+01  5.308943712E-03  1.733339616E+01 -6.168190930E-03
2.0   10.0   2.281518968E+03  2.150981701E-05  2.214684510E+03 -2.295073686E-05
2.0   20.0   3.931278522E+07  6.329543612E-10  3.852369486E+07 -6.516012331E-10
3.0   1.0    2.216842492E-02  7.101262825E+00  6.924239499E-02 -2.292862737E+01
3.0   2.0    2.127399592E-01  6.473853909E-01  3.698385088E-01 -1.224837841E+00
3.0   5.0    1.033115017E+01  8.291768415E-03  1.130692487E+01 -1.028400476E-02
3.0   10.0   1.758380717E+03  2.725270026E-05  1.754004753E+03 -2.968562708E-05
3.0   50.0   2.677764139E+20  3.727936774E-23  2.655764792E+20 -3.771608045E-23
5.0   1.0    2.714631560E-04  3.609605896E+02  1.379804441E-03 -1.849035364E+03
5.0   2.0    9.825679323E-03  9.431049101E+00  2.616437167E-02 -2.577353868E+01
5.0   5.0    2.157974547E+00  3.270627371E-02  2.950260216E+00 -4.796533952E-02
5.0   10.0   7.771882864E+02  5.754184999E-05  8.378963946E+02 -6.663236215E-05
```

```
 5.0   50.0   2.278548308E+20  4.367182254E-23  2.267244113E+20 -4.432002477E-23
10.0    1.0   2.752948040E-10  1.807132899E+08  2.765437823E-09 -1.817137940E+09
10.0    2.0   3.016963879E-07  1.624824040E+05  1.535703963E-06 -8.302224962E+05
10.0    5.0   4.580044419E-03  9.758562829E+00  1.015562998E-02 -2.202940355E+01
10.0   10.0   2.189170616E+01  1.614255300E-03  3.042758634E+01 -2.324264134E-03
10.0   50.0   1.071597159E+20  9.150988210E-23  1.082473779E+20 -9.419859989E-23
20.0    1.0   3.966835986E-25  6.294369360E+22  7.943111714E-24 -1.260529076E+24
20.0    2.0   4.310560576E-19  5.770856853E+16  4.331042808E-18 -5.801141519E+17
20.0    5.0   5.024239358E-11  4.827000521E+08  2.068719274E-10 -1.993195442E+09
20.0   10.0   1.250799736E-04  1.787442782E+02  2.784778118E-04 -4.015325804E+02
20.0   50.0   5.442008403E+18  1.706148380E-21  5.814242572E+18 -1.852264589E-21
Spherical Bessel Functions        n,x,sjn(x),syn(x),sjnp(x),synp(x)
11 Values
 0    0.1    9.9833417E-01  -9.9500417E+00  -3.3300012E-02   1.0049875E+02
 0    1.0    8.4147098E-01  -5.4030231E-01  -3.0116868E-01   1.3817733E+00
 0    5.0   -1.9178485E-01  -5.6732437E-02   9.5089408E-02  -1.8043837E-01
 0   50.0   -5.2474971E-03  -1.9299321E-02   1.9404271E-02  -4.8615107E-03
 1    0.1    3.3300012E-02  -1.0049875E+02   3.3233393E-01   2.0000250E+03
 1    1.0    3.0116868E-01  -1.3817733E+00   2.3913363E-01   2.2232443E+00
 1    5.0   -9.5089408E-02   1.8043837E-01  -1.5374909E-01  -1.2890778E-01
 1   50.0   -1.9404271E-02   4.8615107E-03  -4.4713263E-03  -1.9493781E-02
20    1.0    7.5377957E-26  -3.2395922E+23   1.5058053E-24   6.7948312E+24
20    5.0    5.4277268E-12  -9.2679514E+08   2.1071414E-11   3.7715819E+09
20   50.0   -1.5785030E-02   1.3759531E-02  -1.2204181E-02  -1.4702296E-02
Airy Functions
8 Values
-3.5  -0.37553382   0.16893984  -0.34344343  -0.69311628
-2.0   0.22740743  -0.41230259   0.61825902   0.27879517
-1.0   0.53556088   0.10399739  -0.01016057   0.59237563
-0.01  0.35761619   0.61044364  -0.25880157   0.44831896
 0.00  0.35502805   0.61492663  -0.25881940   0.44828836
 0.01  0.35243992   0.61940962  -0.25880174   0.44831926
 0.5   0.23169361   0.85427704  -0.22491053   0.54457256
 1.00  0.13529242   1.20742359  -0.15914744   0.93243593
Elliptic Integral First Kind RF
3 Values
  0.1     0.1     0.1     3.16227766
  0.0     1.0     2.0     1.31102878
100.0   100.0   100.0     0.1
Elliptic Integral Second Kind RD
3 Values
  0.1     0.1     0.1    31.6227766
  0.0     2.0     1.0     1.79721036
100.0   100.0   100.0     0.001
Elliptic Integral Third Kind RJ
3 Values
  0.1     0.1     0.1     0.1    31.6227766
  2.0     3.0     4.0     5.0     0.142975797
100.0   100.0   100.0   100.0     0.001
Elliptic Integral Degenerate RC
5 Values
  0.1     0.1     3.16227766
  0.0     0.25    3.14159265
  0.0625  0.125   3.14159265
  2.25    2.0     0.69314718
100.0   100.0     0.1
Legendre Elliptic Integral First Kind
```

```
9 Values
   5.0    2.0     0.08726660
   5.0   30.0     0.08729413
   5.0   88.0     0.08737730
  30.0    2.0     0.52362636
  30.0   30.0     0.52942863
  30.0   88.0     0.54927042
  90.0    2.0     1.57127495
  90.0   30.0     1.68575035
  90.0   88.0     4.74271727
Legendre Elliptic Integral Second Kind
9 Values
   5.0    2.0     0.08726633
   5.0   30.0     0.08723881
   5.0   88.0     0.08715588
  30.0    2.0     0.52357119
  30.0   30.0     0.51788193
  30.0   88.0     0.50003003
  90.0    2.0     1.57031792
  90.0   30.0     1.46746221
  90.0   88.0     1.00258409
Legendre Elliptic Integral Third Kind
12 Values
  15.0    0.4    30.0     0.264953
  60.0    0.6    90.0     1.71951
  90.0    0.0    30.0     1.68575
  90.0    0.9    75.0    12.46407
  15.0    1.0    15.0     0.268156
  45.0    0.0625 30.0     0.813845
  45.0,   0.625  30.0     0.921130
  45.0    1.25   30.0     1.132136
  45.0   -0.25   30.0     0.769872
  90.0    0.0625 30.0     1.743055
  90.0    0.625  30.0     2.800989
  90.0   -0.25   30.0     1.501762
Fresnel Integrals
6 Values
   0.1    0.0005236    0.0999975
   1.0    0.4382591    0.7798934
   1.5    0.6975050    0.4452612
   2.0    0.3434157    0.4882534
   5.0    0.4991914    0.5636312
  20.0    0.4840845    0.4999873
Exponential Integral En        n,x,En(x)
13 Values
   0      1.0     0.3678794
   2      0.0     1.0000000
   3      0.0     0.5000000
   4      0.0     0.3333333
   2      0.5     0.3266439
   3      0.5     0.2216044
   4      0.5     0.1652428
  10      0.5     0.0634583
  20      0.5     0.0310612
   2      5.0     0.9964690E-03
  20      5.0     0.2782746E-03
   2     50.0     0.3711783E-23
```

```
20    50.0      0.2766423E-23
Cosine and Sine Integrals
17 Values
  0.1   -1.727868E+00     9.994446E-02
  0.2   -1.042206E+00     1.995561E-01
  0.6   -2.227071E-02     5.881288E-01
  0.7    1.005147E-01     6.812222E-01
  1.8    4.568111E-01     1.505817E+00
  1.9    4.419403E-01     1.557775E+00
  2.0    4.229808E-01     1.605413E+00
  2.1    4.005120E-01     1.648699E+00
  2.2    3.750746E-01     1.687625E+00
  3.3    2.467829E-02     1.848081E+00
  6.3   -1.988822E-02     1.418174E+00
  6.4   -4.181411E-03     1.419223E+00
  6.5    1.110152E-02     1.421794E+00
  6.6    2.582314E-02     1.425816E+00
 10.0   -4.545644E-02     1.658348E+00
 12.5   -1.140835E-02     1.492337E+00
 15.0    4.627868E-02     1.618194E+00
Exponential Integral Ei
21 Values
  0.1   -1.62281
  0.2   -0.821761
  0.3   -0.302669
  0.4    0.104768
  0.5    0.454220
  0.6    0.769881
  0.7    1.06491
  0.8    1.34740
  0.9    1.62281
  2.2    5.73261
  2.4    6.60067
  3.5    13.9254
  3.9    18.3157
  5.0    40.1853
  5.4    54.1935
 12.5    23565.1
 13.0    37197.7
 13.5    58827.0
 14.0    93193.0
 14.5    147866.
 15.0    234955.
Gamma Function
17 Values
  1.0    1.000000
  1.2    0.918169
  1.4    0.887264
  1.6    0.893515
  1.8    0.931384
  2.0    1.000000
  0.2    4.590845
  0.4    2.218160
  0.6    1.489192
  0.8    1.164230
 -0.2    5.2005665E01
 -0.4    4.617091E01
```

```
-0.6    4.0128959E01
-0.8    3.4231564E01
10.0    3.6288000E05
20.0    1.2164510E17
30.0    8.8417620E30
N-factorial
18 Values
1       1
2       2
3       6
4       24
5       120
6       720
7       5040
8       40320
9       362880
10      3628800
11      39916800
12      479001600
13      6227020800
14      87178291200
15      1.3076755E12
20      2.4329042E18
25      1.5511222E25
30      2.6525281E32
Binomial Coefficients
20 Values
1       0       1
6       1       6
6       3       20
6       5       6
15      1       15
15      3       455
15      5       3003
15      7       6435
15      9       5005
15      11      1365
15      13      105
25      1       25
25      3       2300
25      5       53130
25      7       480700
25      9       2042975
25      11      4457400
25      13      5200300
25      15      3268760
25      17      1081575
Beta Function
15 Values
1.0     1.0     1.000000
0.2     1.0     5.000000
1.0     0.2     5.000000
0.4     1.0     2.500000
1.0     0.4     2.500000
0.6     1.0     1.666667
0.8     1.0     1.250000
6.0     6.0     3.607504E-04
```

```
6.0    5.0    7.936508E-04
6.0    4.0    1.984127E-03
6.0    3.0    5.952381E-03
6.0    2.0    0.238095E-01
7.0    7.0    8.325008E-05
5.0    5.0    1.587302E-03
4.0    4.0    7.142857E-03
3.0    3.0    0.333333E-01
2.0    2.0    1.666667E-01
```

Incomplete Gamma Function
20 Values
```
0.1    3.1622777E-02    0.7420263
0.1    3.1622777E-01    0.9119753
0.1    1.5811388        0.9898955
0.5    7.0710678E-02    0.2931279
0.5    7.0710678E-01    0.7656418
0.5    3.5355339        0.9921661
1.0    0.1000000        0.0951626
1.0    1.0000000        0.6321206
1.0    5.0000000        0.9932621
1.1    1.0488088E-01    0.0757471
1.1    1.0488088        0.6076457
1.1    5.2440442        0.9933425
2.0    1.4142136E-01    0.0091054
2.0    1.4142136        0.4130643
2.0    7.0710678        0.9931450
6.0    2.4494897        0.0387318
6.0    12.247449        0.9825937
11.0   16.583124        0.9404267
26.0   25.495098        0.4863866
41.0   44.821870        0.7359709
```

Error Function
20 Values
```
0.0    0.000000
0.1    0.1124629
0.2    0.2227026
0.3    0.3286268
0.4    0.4283924
0.5    0.5204999
0.6    0.6038561
0.7    0.6778012
0.8    0.7421010
0.9    0.7969082
1.0    0.8427008
1.1    0.8802051
1.2    0.9103140
1.3    0.9340079
1.4    0.9522851
1.5    0.9661051
1.6    0.9763484
1.7    0.9837905
1.8    0.9890905
1.9    0.9927904
```

Incomplete Beta Function
20 Values
```
0.5    0.5    0.01    0.0637686
0.5    0.5    0.10    0.2048328
```

0.5	0.5	1.00	1.0000000
1.0	0.5	0.01	0.0050126
1.0	0.5	0.10	0.0513167
1.0	0.5	1.00	1.0000000
1.0	1.0	0.5	0.5000000
5.0	5.0	0.5	0.5000000
10.0	0.5	0.9	0.1516409
10.0	5.0	0.5	0.0897827
10.0	5.0	1.0	1.0000000
10.0	10.0	0.5	0.5000000
20.0	5.0	0.8	0.4598773
20.0	10.0	0.6	0.2146816
20.0	10.0	0.8	0.9507365
20.0	20.0	0.5	0.5000000
20.0	20.0	0.6	0.8979414
30.0	10.0	0.7	0.2241297
30.0	10.0	0.8	0.7586405
40.0	20.0	0.7	0.7001783

Bessel Function J0
20 Values

-5.0	-0.1775968
-4.0	-0.3971498
-3.0	-0.2600520
-2.0	0.2238908
-1.0	0.7651976
0.0	1.0000000
1.0	0.7651977
2.0	0.2238908
3.0	-0.2600520
4.0	-0.3971498
5.0	-0.1775968
6.0	0.1506453
7.0	0.3000793
8.0	0.1716508
9.0	-0.0903336
10.0	-0.2459358
11.0	-0.1711903
12.0	0.0476893
13.0	0.2069261
14.0	0.1710735
15.0	-0.0142245

Bessel Function Y0
15 Values

0.1	-1.5342387
1.0	0.0882570
2.0	0.51037567
3.0	0.37685001
4.0	-0.0169407
5.0	-0.3085176
6.0	-0.2881947
7.0	-0.0259497
8.0	0.2235215
9.0	0.2499367
10.0	0.0556712
11.0	-0.1688473
12.0	-0.2252373
13.0	-0.0782079

```
14.0    0.1271926
15.0    0.2054743
Bessel Function J1
20 Values
-5.0    0.3275791
-4.0    0.0660433
-3.0   -0.3390590
-2.0   -0.5767248
-1.0   -0.4400506
 0.0    0.0000000
 1.0    0.4400506
 2.0    0.5767248
 3.0    0.3390590
 4.0   -0.0660433
 5.0   -0.3275791
 6.0   -0.2766839
 7.0   -0.0046828
 8.0    0.2346364
 9.0    0.2453118
10.0    0.0434728
11.0   -0.1767853
12.0   -0.2234471
13.0   -0.0703181
14.0    0.1333752
15.0    0.2051040
Bessel Function Y1
15 Values
 0.1   -6.4589511
 1.0   -0.7812128
 2.0   -0.1070324
 3.0    0.3246744
 4.0    0.3979257
 5.0    0.1478631
 6.0   -0.1750103
 7.0   -0.3026672
 8.0   -0.1580605
 9.0    0.1043146
10.0    0.2490154
11.0    0.1637055
12.0   -0.0570992
13.0   -0.2100814
14.0   -0.1666448
15.0    0.0210736
Bessel Function Jn, n>=2
20 Values
2      1.0     1.149034849E-01
2      2.0     3.528340286E-01
2      5.0     4.656511628E-02
2     10.0     2.546303137E-01
2     50.0    -5.971280079E-02
5      1.0     2.497577302E-04
5      2.0     7.039629756E-03
5      5.0     2.611405461E-01
5     10.0    -2.340615282E-01
5     50.0    -8.140024770E-02
10     1.0     2.630615124E-10
10     2.0     2.515386283E-07
```

```
10      5.0     1.467802647E-03
10      10.0    2.074861066E-01
10      50.0   -1.138478491E-01
20      1.0     3.873503009E-25
20      2.0     3.918972805E-19
20      5.0     2.770330052E-11
20      10.0    1.151336925E-05
20      50.0   -1.167043528E-01
Bessel Function Yn, n>=2
20 Values
2       1.0    -1.650682607
2       2.0    -6.174081042E-01
2       5.0     3.676628826E-01
2       10.0   -5.868082460E-03
2       50.0    9.579316873E-02
5       1.0    -2.604058666E02
5       2.0    -9.935989128
5       5.0    -4.536948225E-01
5       10.0    1.354030477E-01
5       50.0   -7.854841391E-02
10      1.0    -1.216180143E08
10      2.0    -1.291845422E05
10      5.0    -2.512911010E01
10      10.0   -3.598141522E-01
10      50.0    5.723897182E-03
20      1.0    -4.113970315E22
20      2.0    -4.081651389E16
20      5.0    -5.933965297E08
20      10.0   -1.597483848E03
20      50.0    1.644263395E-02
Modified Bessel Function I0
20 Values
0.0     1.0000000
0.2     1.0100250
0.4     1.0404018
0.6     1.0920453
0.8     1.1665149
1.0     1.2660658
1.2     1.3937256
1.4     1.5533951
1.6     1.7499807
1.8     1.9895593
2.0     2.2795852
2.5     3.2898391
3.0     4.8807925
3.5     7.3782035
4.0     11.301922
4.5     17.481172
5.0     27.239871
6.0     67.234406
8.0     427.56411
10.0    2815.7167
Modified Bessel Function K0
20 Values
0.1     2.4270690
0.2     1.7527038
0.4     1.1145291
```

```
0.6     0.77752208
0.8     0.56534710
1.0     0.42102445
1.2     0.31850821
1.4     0.24365506
1.6     0.18795475
1.8     0.14593140
2.0     0.11389387
2.5     6.2347553E-02
3.0     3.4739500E-02
3.5     1.9598897E-02
4.0     1.1159676E-02
4.5     6.3998572E-03
5.0     3.6910983E-03
6.0     1.2439943E-03
8.0     1.4647071E-04
10.0    1.7780062E-05
```
Modified Bessel Function I1
20 Values
```
0.0     0.00000000
0.2     0.10050083
0.4     0.20402675
0.6     0.31370403
0.8     0.43286480
1.0     0.56515912
1.2     0.71467794
1.4     0.88609197
1.6     1.0848107
1.8     1.3171674
2.0     1.5906369
2.5     2.5167163
3.0     3.9533700
3.5     6.2058350
4.0     9.7594652
4.5     15.389221
5.0     24.335643
6.0     61.341937
8.0     399.87313
10.0    2670.9883
```
Modified Bessel Function K1
20 Values
```
0.1     9.8538451
0.2     4.7759725
0.4     2.1843544
0.6     1.3028349
0.8     0.86178163
1.0     0.60190724
1.2     0.43459241
1.4     0.32083589
1.6     0.24063392
1.8     0.18262309
2.0     0.13986588
2.5     7.3890816E-02
3.0     4.0156431E-02
3.5     2.2239393E-02
4.0     1.2483499E-02
4.5     7.0780949E-03
```

```
5.0    4.0446134E-03
6.0    1.3439197E-03
8.0    1.5536921E-04
10.0   1.8648773E-05
```
Modified Bessel Function Kn, n>=2
28 Values

```
2     0.2     49.512430
2     1.0     1.6248389
2     2.0     2.5375975E-01
2     2.5     1.2146021E-01
2     3.0     6.1510459E-02
2     5.0     5.3089437E-03
2    10.0     2.1509817E-05
2    20.0     6.3295437E-10
3     1.0     7.101262825
3     2.0     6.473853909E-01
3     5.0     8.291768415E-03
3    10.0     2.725270026E-05
3    50.0     3.72793677E-23
5     1.0     3.609605896E02
5     2.0     9.431049101
5     5.0     3.270627371E-02
5    10.0     5.754184999E-05
5    50.0     4.36718224E-23
10    1.0     1.807132899E08
10    2.0     1.624824040E05
10    5.0     9.758562829
10   10.0     1.614255300E-03
10   50.0     9.15098819E-23
20    1.0     6.294369369E22
20    2.0     5.770856853E16
20    5.0     4.827000521E08
20   10.0     1.787442782E02
20   50.0     1.70614838E-21
```
Modified Bessel Function In, n>=2
28 Values

```
2     0.2     5.0166876E-03
2     1.0     1.3574767E-01
2     2.0     6.8894844E-01
2     2.5     1.2764661
2     3.0     2.2452125
2     5.0     17.505615
2    10.0     2281.5189
2    20.0     3.9312785E07
3     1.0     2.216842492E-02
3     2.0     2.127399592E-01
3     5.0     1.033115017E01
3    10.0     1.758380717E01
3    50.0     2.67776414E20
5     1.0     2.714631560E-04
5     2.0     9.825679323E-03
5     5.0     2.157974547
5    10.0     7.771882864E02
5    50.0     2.27854831E20
10    1.0     2.752948040E-10
10    2.0     3.016963879E-07
10    5.0     4.580044419E-03
```

```
10     10.0    2.189170616E01
10     50.0    1.07159716E20
20     1.0     3.966835986E-25
20     2.0     4.310560576E-19
20     5.0     5.024239358E-11
20     10.0    1.250799736E-04
20     50.0    5.44200840E18
```

Legendre Polynomials
19 Values

```
1     0     1.0          1.224745
10    0     1.0          3.240370
20    0     1.0          4.527693
1     0     0.7071067    0.866025
10    0     0.7071067    0.373006
20    0     0.7071067   -0.874140
1     0     0.0          0.000000
10    0     0.0         -0.797435
20    0     0.0          0.797766
2     2     0.7071067    0.484123
10    2     0.7071067   -0.204789
20    2     0.7071067    0.910208
2     2     0.0          0.968246
10    2     0.0          0.804785
20    2     0.0         -0.799672
10    10    0.7071067    0.042505
20    10    0.7071067   -0.707252
10    10    0.0          1.360172
20    10    0.0         -0.853705
```

Jacobian Elliptic Function
20 Values

```
0.0    0.1     0.099833
0.0    0.2     0.19867
0.0    0.5     0.47943
0.0    1.0     0.84147
0.0    2.0     0.90930
0.5    0.1     0.099751
0.5    0.2     0.19802
0.5    0.5     0.47075
0.5    1.0     0.80300
0.5    2.0     0.99466
1.0    0.1     0.099668
1.0    0.2     0.19738
1.0    0.5     0.46212
1.0    1.0     0.76159
1.0    2.0     0.96403
1.0    4.0     0.99933
1.0   -0.2    -0.19738
1.0   -0.5    -0.46212
1.0   -1.0    -0.76159
1.0   -2.0    -0.96403
```

Chapter 7: Random Numbers

Chapter 7 of Numerical Recipes deals with the generation of random numbers drawn from various distributions. The first four subroutines produce uniform deviates with a range of 0.0 to 1.0. ran0 *is a subroutine that implements the Park and Miller "minimal standard" random number generator.* ran1 *adds a shuffle to the minimal standard.* ran2 *obtains deviates by the L'Ecuyer long period algorithm plus a shuffle.* ran3 *is Knuth's portable generator, based on a subtractive rather than a congruential method.*

The transformation method is used to generate some non-uniform distributions. Resulting from this are routines expdev, *which gives exponentially distributed deviates, and* gasdev *for Gaussian deviates. The rejection method of producing non-uniform deviates is also discussed, and is used in* gamdev *(deviate with a gamma function distribution),* poidev *(deviate with a Poisson distribution), and* bnldev *(deviate with a binomial distribution). For generating random sequences of zeros and ones, there are two subroutines,* irbit1 *and* irbit2, *both based on the 18 lowest significant bits in the seed* iseed, *but each using a different recurrence to proceed from step to step.*

The "pseudo-DES" (Data Encryption Standard) hashing of 64-bit words is carried out by psdes. *It is used to generate random deviates by* ran4. sobseq *produces Sobol's quasi-random sequence. The routines* vegas *and* miser *are used for multidimensional Monte-Carlo integration, with different strategies for picking points in n-dimensional space.*

$$\star \quad \star \quad \star \quad \star$$

Sample program xran tests each random number generator, ran0 to ran3, in turn. For each one, it first draws four consecutive random numbers $X_1, \ldots, X_4$ from the generator in question. Then it treats the numbers as coordinates of a point. For example, it takes (X_1, X_2) as a point in two dimensions, (X_1, X_2, X_3) as a point in three dimensions, etc. These points are inside boxes of unit dimension in their respective n-space. They may, however, be either inside or outside of the unit sphere in that space. For $n = 2, 3, 4$ we seek the probability that a point is inside the unit n-sphere. This number is easily calculated theoretically. For $n = 2$ it is $\pi/4$, for $n = 3$ it is $\pi/6$, and for $n = 4$ it is $\pi^2/32$. If the random number generator is not faulty, the points will fall within the unit n-sphere this fraction of the time, and the result should become increasingly accurate as the number of points increases. In the program xran we have taken out a factor of 2^n for convenience, and used the random number generators as a statistical means of determining the value of π, $4\pi/3$, and $\pi^2/2$.

```
      PROGRAM xran
C     driver for routines ran0, ran1, ran2, ran3
      REAL ran0,ran1,ran2,ran3
      EXTERNAL ran0,ran1,ran2,ran3
      write(*,*)
      write(*,*) 'Testing ran0:'
      call integ(ran0)
      write(*,*)
      write(*,*) 'Testing ran1:'
      call integ(ran1)
      write(*,*)
      write(*,*) 'Testing ran2:'
      call integ(ran2)
      write(*,*)
      write(*,*) 'Testing ran3:'
      call integ(ran3)
      END

      SUBROUTINE integ(func)
C     calculates pi statistically using volume of unit n-sphere
      REAL PI
      PARAMETER(PI=3.1415926)
      INTEGER i,idum,j,k,iy(3)
      REAL fnc,func,x1,x2,x3,x4,yprob(3)
      EXTERNAL func
      fnc(x1,x2,x3,x4)=sqrt(x1**2+x2**2+x3**2+x4**2)
      idum=-1
      do 11 i=1,3
        iy(i)=0
11    continue
      write(*,'(1x,t15,a)') 'Volume of unit n-sphere, n = 2, 3, 4'
      write(*,'(1x,/,t3,a,t17,a,t26,a,t37,a)')
     *     '# points','PI','(4/3)*PI','(1/2)*PI^2'
      do 14 j=1,15
        do 12 k=2**(j-1),0,-1
          x1=func(idum)
          x2=func(idum)
          x3=func(idum)
          x4=func(idum)
          if (fnc(x1,x2,0.0,0.0).lt.1.0) iy(1)=iy(1)+1
          if (fnc(x1,x2,x3,0.0).lt.1.0) iy(2)=iy(2)+1
          if (fnc(x1,x2,x3,x4).lt.1.0) iy(3)=iy(3)+1
12      continue
        do 13 i=1,3
          yprob(i)=1.0*(2**(i+1))*iy(i)/(2**j)
13      continue
        write(*,'(1x,i8,3f12.6)') 2**j,(yprob(i),i=1,3)
14    continue
      write(*,'(1x,/,t4,a,3f12.6,/)') 'actual',PI,4.0*PI/3.0,0.5*(PI**2)
      return
      END
```

Routine `expdev` generates random numbers drawn from an exponential deviate. Sample program `xexpdev` makes ten thousand calls to `expdev` and bins the results into 21 bins, the contents of which are tallied in array `x(i)`. Then the sum `total` of all bins is taken, since some of the numbers will be too large to have fallen in any of the bins. The `x(i)` are scaled to `total`, and then compared to a similarly normalized exponential which is called `expect`.

```
      PROGRAM xexpdev
C     driver for routine expdev
      INTEGER NPTS
      REAL EE
      PARAMETER(NPTS=10000,EE=2.718281828)
      INTEGER i,idum,j
      REAL expdev,expect,total,y,trig(21),x(21)
      do 11 i=1,21
        trig(i)=(i-1)/20.0
        x(i)=0.0
11    continue
      idum=-1
      do 13 i=1,NPTS
        y=expdev(idum)
        do 12 j=2,21
          if ((y.lt.trig(j)).and.(y.gt.trig(j-1))) then
            x(j)=x(j)+1.0
          endif
12      continue
13    continue
      total=0.0
      do 14 i=2,21
        total=total+x(i)
14    continue
      write(*,'(1x,a,i6,a)') 'Exponential distribution with',
     *    NPTS,' points:'
      write(*,'(1x,t5,a,t19,a,t31,a)')
     *    'interval','observed','expected'
      do 15 i=2,21
        x(i)=x(i)/total
        expect=exp(-(trig(i-1)+trig(i))/2.0)
        expect=expect*0.05*EE/(EE-1)
        write(*,'(1x,2f6.2,2f12.4)')
     *        trig(i-1),trig(i),x(i),expect
15    continue
      END
```

`gasdev` generates random numbers from a Gaussian deviate. Example `xgasdev` takes ten thousand of these and puts them into 21 bins. For the purpose of binning, the center of the Gaussian is shifted over by `NOVER2=10` bins, to put it in the middle bin. The remainder of the program simply plots the contents of the bins, to illustrate that they have the characteristic Gaussian bell-shape. This allows a quick, though superficial, check of the integrity of the routine.

```
      PROGRAM xgasdev
C     driver for routine gasdev
      INTEGER N,NP1,NOVER2,NPTS,ISCAL,LLEN
      PARAMETER(N=20,NP1=N+1,NOVER2=N/2,NPTS=10000,ISCAL=400,LLEN=50)
      INTEGER i,idum,j,k,klim
      REAL gasdev,x,dist(NP1)
      CHARACTER text(50)*1
      idum=-13
      do 11 j=1,NP1
        dist(j)=0.0
```

```
11      continue
        do 12 i=1,NPTS
          j=nint(0.25*N*gasdev(idum))+NOVER2+1
          if ((j.ge.1).and.(j.le.NP1)) dist(j)=dist(j)+1
12      continue
        write(*,'(1x,a,i6,a)')
     *      'Normally distributed deviate of ',NPTS,' points'
        write(*,'(1x,t6,a,t14,a,t23,a)') 'x','p(x)','graph:'
        do 15 j=1,NP1
          dist(j)=dist(j)/NPTS
          do 13 k=1,50
            text(k)=' '
13        continue
          klim=int(ISCAL*dist(j))
          if (klim.gt.LLEN) klim=LLEN
          do 14 k=1,klim
            text(k)='*'
14        continue
          x=float(j)/(0.25*N)
          write(*,'(1x,f7.2,f10.4,4x,50a1)')
     *        x,dist(j),(text(k),k=1,50)
15      continue
        END
```

The next three sample programs, `xgamdev`, `xpoidev` and `xbnldev`, are identical to the previous one, but each drives a different random number generator and produces a different graph. `xgamdev` drives `gamdev` and displays a gamma distribution of order `ia` specified by the user. `xpoidev` drives `poidev` and produces a Poisson distribution with mean `xm` specified by the user. `xbnldev` drives `bnldev` and produces a binomial distribution, also with specified `xm`.

```
        PROGRAM xgamdev
C       driver for routine gamdev
        INTEGER N,NPTS,ISCAL,LLEN
        PARAMETER(N=20,NPTS=10000,ISCAL=200,LLEN=50)
        INTEGER i,ia,idum,j,k,klim
        REAL gamdev,dist(21)
        CHARACTER text(50)*1
        idum=-13
10      do 11 j=1,21
          dist(j)=0.0
11      continue
        write(*,*) 'Order of Gamma distribution (n=1..20); -1 to END.'
        read(*,*) ia
        if (ia.le.0) goto 99
        if (ia.gt.20) goto 10
        do 12 i=1,NPTS
          j=int(gamdev(ia,idum))+1
          if ((j.ge.1).and.(j.le.21)) dist(j)=dist(j)+1
12      continue
        write(*,'(1x,a,i2,a,i6,a)') 'Gamma-distribution deviate, order ',
     *      ia,' of ',NPTS,' points'
        write(*,'(1x,t6,a,t14,a,t23,a)') 'x','p(x)','graph:'
        do 15 j=1,20
          dist(j)=dist(j)/NPTS
          do 13 k=1,50
```

```
            text(k)=' '
13       continue
         klim=int(ISCAL*dist(j))
         if (klim.gt.LLEN) klim=LLEN
         do 14 k=1,klim
            text(k)='*'
14       continue
         write(*,'(1x,f7.2,f10.4,4x,50a1)')
     *        float(j),dist(j),(text(k),k=1,50)
15    continue
      goto 10
99    END


      PROGRAM xpoidev
C     driver for routine poidev
      INTEGER N,NPTS,ISCAL,LLEN
      PARAMETER(N=20,NPTS=10000,ISCAL=200,LLEN=50)
      INTEGER i,idum,j,k,klim
      REAL poidev,xm,dist(21)
      CHARACTER text(50)*1
      idum=-13
10    do 11 j=1,21
         dist(j)=0.0
11    continue
      write(*,*) 'Mean of Poisson distrib. (x=0 to 20); neg. to END'
      read(*,*) xm
      if (xm.lt.0.0) goto 99
      if (xm.gt.20.0) goto 10
      do 12 i=1,NPTS
         j=int(poidev(xm,idum))+1
         if ((j.ge.1).and.(j.le.21)) dist(j)=dist(j)+1
12    continue
      write(*,'(1x,a,f5.2,a,i6,a)')
     *     'Poisson-distributed deviate, mean ',
     *     xm,' of ',NPTS,' points'
      write(*,'(1x,t6,a,t14,a,t23,a)') 'x','p(x)','graph:'
      do 15 j=1,20
         dist(j)=dist(j)/NPTS
         do 13 k=1,50
            text(k)=' '
13       continue
         klim=int(ISCAL*dist(j))
         if (klim.gt.LLEN) klim=LLEN
         do 14 k=1,klim
            text(k)='*'
14       continue
         write(*,'(1x,f7.2,f10.4,4x,50a1)')
     *        float(j),dist(j),(text(k),k=1,50)
15    continue
      goto 10
99    END


      PROGRAM xbnldev
C     driver for routine bnldev
      INTEGER N,NPTS,ISCAL,NN
      PARAMETER(N=20,NPTS=1000,ISCAL=200,NN=100)
      INTEGER i,idum,j,k,klim,llen
```

```
      REAL bnldev,dist(21),pp,xm
      CHARACTER text(50)*1
      idum=-133
      llen=50
10    do 11 j=1,21
        dist(j)=0.0
11    continue
      write(*,*)
     *    'Mean of binomial distribution (0 to 20) (Negative to END)'
      read(*,*) xm
      if (xm.lt.0) goto 99
      pp=xm/NN
      do 12 i=1,NPTS
        j=int(bnldev(pp,NN,idum))
        if ((j.ge.0).and.(j.le.20)) dist(j+1)=dist(j+1)+1
12    continue
      write(*,'(1x,t5,a,t10,a,t18,a)') 'x','p(x)','graph:'
      do 15 j=1,20
        dist(j)=dist(j)/NPTS
        do 13 k=1,50
          text(k)=' '
13      continue
        text(1)='*'
        klim=int(ISCAL*dist(j))
        if (klim.gt.llen) klim=llen
        do 14 k=1,klim
          text(k)='*'
14      continue
        write(*,'(1x,f5.1,f8.4,3x,50a1)') float(j-1),dist(j),
     *      (text(k),k=1,50)
15    continue
      goto 10
99    END
```

Subroutines `irbit1` and `irbit2` both generate random series of ones and zeros. The sample programs `xirbit1` and `xirbit2` for the two are the same, and they check that the series have correct statistical properties (or more exactly, that they have at least one correct property). They look for a 1 in the series and count how many zeros follow it before the next 1 appears. The result is stored as a distribution. There should be, for example, a 50% chance of no zeros, a 25% chance of exactly one zero, and so on.

```
      PROGRAM xirbit1
C     driver for routine irbit1
      INTEGER NBIN,NTRIES
      PARAMETER(NBIN=15,NTRIES=10000)
      INTEGER i,idum,iflg,ipts,irbit1,iseed,j,n
      REAL delay(NBIN)
      iseed=12345
      do 11 i=1,NBIN
        delay(i)=0.0
11    continue
      ipts=0
      do 13 i=1,NTRIES
        if (irbit1(iseed).eq.1) then
          ipts=ipts+1
```

```
        iflg=0
        do 12 j=1,NBIN
          idum=irbit1(iseed)
          if ((idum.eq.1).and.(iflg.eq.0)) then
            iflg=1
            delay(j)=delay(j)+1.0
          endif
12        continue
        endif
13    continue
      write(*,*) 'Distribution of runs of N zeros'
      write(*,'(1x,t7,a,t16,a,t38,a)') 'N','Probability','Expected'
      do 14 n=1,NBIN
        write(*,'(1x,i6,f18.6,f20.6)')
   *         n-1,delay(n)/ipts,1/(2.0**n)
14    continue
      END


      PROGRAM xirbit2
C     driver for routine irbit2
      INTEGER NBIN,NTRIES
      PARAMETER(NBIN=15,NTRIES=10000)
      INTEGER i,idum,iflg,ipts,irbit2,iseed,j,n
      REAL delay(NBIN)
      iseed=111
      do 11 i=1,NBIN
        delay(i)=0.0
11    continue
      ipts=0
      do 13 i=1,NTRIES
        if (irbit2(iseed).eq.1) then
          ipts=ipts+1
          iflg=0
          do 12 j=1,NBIN
            idum=irbit2(iseed)
            if ((idum.eq.1).and.(iflg.eq.0)) then
              iflg=1
              delay(j)=delay(j)+1.0
            endif
12          continue
        endif
13    continue
      write(*,*) 'Distribution of runs of N zeros'
      write(*,'(1x,t7,a,t16,a,t38,a)') 'N','Probability','Expected'
      do 14 n=1,NBIN
        write(*,'(1x,i6,f18.6,f20.6)')
   *         n-1,delay(n)/ipts,1/(2.0**n)
14    continue
      END
```

The routine psdes uses the logic of the Data Encryption Standard, though with a simplified nonlinear function, to fully "hash" both of its 4-byte arguments. Since the routine makes some assumptions about the overflow behavior of integer arithmetic, it is not strictly portable. On many machines (e.g., VAXes), it needs to be compiled with a special switch, indicating that overflows are *not* to be trapped.

The following test routine simply verifies that the test values suggested in the main book are in fact obtained. Note the need for an additional routine just to print out hexadecimal values. While most compilers have a straightforward format field descriptor like "Z8" to perform this function, it is in fact not part of standard FORTRAN-77. (You might well wonder why we should pay so much attention to the standard when psdes itself is not conformant to it! The answer is that, otherwise, you might not even be able to *determine* whether psdes functions correctly.)

```fortran
      PROGRAM xpsdes
C     driver for routine psdes
      CHARACTER*10 hexout
      INTEGER*4 i,lword(4),irword(4),lans(4),irans(4)
      DATA lword /1,1,99,99/, irword /1,99,1,99/
      DATA lans /1615666638,-645954191,2015506589,-671910160/
      DATA irans /1352404003,-1502825446,1680869764,1505397227/
      do 11 i=1,4
        call psdes(lword(i),irword(i))
        write(*,*) 'PSDES now calculates:      ',
     *      hexout(lword(i)),'  ',hexout(irword(i))
        write(*,*) 'Known correct answers are: ',
     *      hexout(lans(i)),'  ',hexout(irans(i))
11    continue
      END

      FUNCTION hexout(num)
C     utility routine for printing out hexadecimal values
C     (many compilers can do this using the nonstandard "Z" format)
      CHARACTER*10 hexout
      INTEGER NCOMP
      PARAMETER (NCOMP=268435455)
      INTEGER i,j,n,num
      CHARACTER*1 hexit(16)
      SAVE hexit
      DATA hexit /'0','1','2','3','4','5','6','7','8','9',
     *    'a','b','c','d','e','f'/
      n=num
      if (n.lt.0) then
        i=mod(n,16)
        if (i.lt.0) i=16+i
        n=n/16
        n=-n
        if (i.ne.0) then
          n=NCOMP-n
        else
          n=NCOMP+1-n
        endif
      else
        i=mod(n,16)
        n=n/16
      endif
      j=10
      hexout(j:j)=hexit(i+1)
2     if (n.gt.0) then
        i=mod(n,16)
        n=n/16
        j=j-1
```

```
      hexout(j:j)=hexit(i+1)
      goto 2
      endif
      j=j-1
      hexout(j:j)='x'
      j=j-1
      hexout(j:j)='0'
      do 11 i=j-1,1,-1
        hexout(i:i)=' '
11    continue
      return
      END
```

Once `psdes` is known to work, then the only trick in making the encryption-based random number generator `ran4` functional is to be sure that its hexadecimal masking values are compatible with your floating point format. The following program attempts to reproduce the test values given in the main book. If your computer produces *either* the IEEE values, *or* the VAX values, then you very likely have things set up correctly. If not, go back and read the discussion in the main book.

Note that the test program, in resetting the value `idum` between calls, forces `ran4` to access certain values (the 1st and 99th) in two different random sequences (sequence -1 and sequence -99). While this random access is an interesting, and sometimes useful, property, most uses of `ran4` would leave `idum` unchanged between calls after a single initializing call with a negative value.

```
      PROGRAM xran4
C     driver for routine ran4
      INTEGER i,idum(4)
      REAL random(4),ran4,ansiee(4),ansvax(4)
      DATA ansvax /0.275898,0.208204,0.034307,0.838676/
      DATA ansiee /0.219120,0.849246,0.375290,0.457334/
      DATA idum /-1,99,-99,99/
      do 11 i=1,4
        random(i)=ran4(idum(i))
11    continue
      write(*,*) 'ran4 gets values: ',(random(i),i=1,4)
      write(*,*) '    IEEE answers: ',(ansiee(i),i=1,4)
      write(*,*) '     VAX answers: ',(ansvax(i),i=1,4)
      END
```

The routine `sobseq` returns successive values of a multi-dimensional, quasi-random Sobol' sequence. The following example shows how the routine is initialized via a negative argument, and then used to obtain the first 32 values of a 3-dimensional sequence such as might be used to sample points in a 3-dimensional cube. In this example, we do nothing with the values other than to print them out. You can check your output against the following:

0.50000	0.50000	0.50000	1
0.25000	0.75000	0.25000	2
0.75000	0.25000	0.75000	3
0.37500	0.62500	0.12500	4
0.87500	0.12500	0.62500	5
0.12500	0.37500	0.37500	6
0.62500	0.87500	0.87500	7
0.31250	0.31250	0.68750	8

0.81250	0.81250	0.18750	9
0.06250	0.56250	0.93750	10
0.56250	0.06250	0.43750	11
0.18750	0.93750	0.56250	12
0.68750	0.43750	0.06250	13
0.43750	0.18750	0.81250	14
0.93750	0.68750	0.31250	15
0.46875	0.84375	0.40625	16
0.96875	0.34375	0.90625	17
0.21875	0.09375	0.15625	18
0.71875	0.59375	0.65625	19
0.09375	0.46875	0.28125	20
0.59375	0.96875	0.78125	21
0.34375	0.71875	0.03125	22
0.84375	0.21875	0.53125	23
0.15625	0.53125	0.84375	24
0.65625	0.03125	0.34375	25
0.40625	0.28125	0.59375	26
0.90625	0.78125	0.09375	27
0.28125	0.15625	0.96875	28
0.78125	0.65625	0.46875	29
0.03125	0.90625	0.71875	30
0.53125	0.40625	0.21875	31
0.26562	0.60938	0.57812	32

```
      PROGRAM xsobseq
C     driver for routine sobseq
      INTEGER j,jj
      REAL x(3)
      call sobseq(-1,x)
      do 11 jj=1,32
        call sobseq(3,x)
        write(*,'(3(1x,f10.5),1x,i5)') (x(j),j=1,3),jj
11    continue
      END
```

We demonstrate the adaptive Monte Carlo integration routine vegas on a narrowly peaked, n-dimensional Gaussian function,

$$f(\mathbf{x}) = \left(\frac{10}{\sqrt{\pi}}\right) e^{-100(\mathbf{x}-\mathbf{x}_0)^2}$$

The function is normalized so that its integral over all space is unity, for any number of dimensions n.

After prompting for a (negative) seed for the random number generator, the demonstration program prompts for several further quantities: First, the number of dimensions. Second, a scalar offset x_0 which is used to construct the vector $\mathbf{x}_0$ as

$$\mathbf{x}_0 = (x_0, x_0, \ldots, x_0)$$

Third, the maximum number of evaluations of (calls to) the function that are allowed on each adaptive iteration. Fourth, the number of adaptive iterations. Fifth, a code that controls how much information is printed out by vegas. The program calls vegas twice, first from a "cold start", and then a second time, inheriting the adapted grid from the first call.

If the five input parameters are entered as

$$4, \quad 0.0, \quad 1000, \quad 4, \quad 0$$

the answer for the integral ought to come out $(1/2)^4 = 0.0625$. Why not 1? Because the offset of 0.0 bisects the Gaussian in each dimension with the boundary of the region.

The following is typical of the actual output produced. Notice that, with this random seed, the first pass gets a very unlucky value at iteration no. 2 (a low value with an unfortunately small error estimate), and the average of all iterations never fully recovers. You would know that something is wrong, however, from the large value (6.1) of the χ^2 per iteration. That is why you might then do a second pass with the now-adapted grid. By the end of the 4th iteration on the second pass, you have a good value, a reliable error estimate, and a credible value for the χ^2.

```
input parameters for vegas:  ndim=  4  ncall=    768.
                             it=   0  itmx=    4
                             nprn=  0  alph= 1.50
                             mds=  1  nd= 50
                             xl( 1)=  0.0000E+00 xu( 1)=   1.000
                             xl( 2)=  0.0000E+00 xu( 2)=   1.000
                             xl( 3)=  0.0000E+00 xu( 3)=   1.000
                             xl( 4)=  0.0000E+00 xu( 4)=   1.000
iteration no.  1: integral = 0.7706086E-01+/-  0.76E-01
all iterations:   integral = 0.7706086E-01+/-  0.76E-01 chi**2/it'n = 0.00E+00
iteration no.  2: integral = 0.2334212E-01+/-  0.75E-02
all iterations:   integral = 0.2386378E-01+/-  0.75E-02 chi**2/it'n = 0.49
iteration no.  3: integral = 0.4258690E-01+/-  0.66E-02
all iterations:   integral = 0.3432380E-01+/-  0.50E-02 chi**2/it'n =  2.0
iteration no.  4: integral = 0.6005644E-01+/-  0.46E-02
all iterations:   integral = 0.4810645E-01+/-  0.34E-02 chi**2/it'n =  6.1
Number of iterations performed:       4
Integral, Standard Dev., Chi-sq.  4.8106447E-02  3.3854251E-03   6.102639
input parameters for vegas:  ndim=  4  ncall=    768.
                             it=   5  itmx=    4
                             nprn=  0  alph= 1.50
                             mds=  1  nd= 50
                             xl( 1)=  0.0000E+00 xu( 1)=   1.000
                             xl( 2)=  0.0000E+00 xu( 2)=   1.000
                             xl( 3)=  0.0000E+00 xu( 3)=   1.000
                             xl( 4)=  0.0000E+00 xu( 4)=   1.000
iteration no.  1: integral = 0.6294765E-01+/-  0.26E-02
all iterations:   integral = 0.6294765E-01+/-  0.26E-02 chi**2/it'n = 0.00E+00
iteration no.  2: integral = 0.6291968E-01+/-  0.20E-02
all iterations:   integral = 0.6293016E-01+/-  0.16E-02 chi**2/it'n = 0.00E+00
iteration no.  3: integral = 0.6065372E-01+/-  0.14E-02
all iterations:   integral = 0.6163474E-01+/-  0.10E-02 chi**2/it'n = 0.58
iteration no.  4: integral = 0.6287680E-01+/-  0.13E-02
all iterations:   integral = 0.6213202E-01+/-  0.81E-03 chi**2/it'n = 0.57
Additional iterations performed:      4
Integral, Standard Dev., Chi-sq.  6.2132023E-02  8.0995343E-04  0.5734860


      PROGRAM xvegas
C     driver for routine vegas
      INTEGER idum,init,itmax,j,ncall,ndim,nprn
      REAL avgi,chi2a,sd,xoff
```

```
      COMMON /ranno/ idum
      COMMON /tofxn/ xoff,ndim
      REAL region(20)
      EXTERNAL fxn
      write(*,*) 'IDUM='
      read(*,*) idum
      idum=-abs(idum)
1     write(*,*) 'ENTER NDIM,XOFF,NCALL,ITMAX,NPRN'
      read(*,*,END=999) ndim,xoff,ncall,itmax,nprn
      avgi=0.
      sd=0.
      chi2a=0.
      do 11 j=1,ndim
        region(j)=0.
        region(j+ndim)=1.
11    continue
      init = -1
      call vegas(region,ndim,fxn,init,ncall,itmax,nprn,avgi,sd,chi2a)
      write(*,*) 'Number of iterations performed:',itmax
      write(*,*) 'Integral, Standard Dev., Chi-sq.',avgi,sd,chi2a
      init = 1
      call vegas(region,ndim,fxn,init,ncall,itmax,nprn,avgi,sd,chi2a)
      write(*,*) 'Additional iterations performed:',itmax
      write(*,*) 'Integral, Standard Dev., Chi-sq.',avgi,sd,chi2a
999   write(*,*) 'NORMAL COMPLETION'
      STOP
      END

      REAL FUNCTION fxn(pt,wgt)
      INTEGER j,ndim
      REAL sum,xoff,wgt
      COMMON /tofxn/ xoff,ndim
      REAL pt(*)
      sum=0.
      do 11 j=1,ndim
        sum=sum+100.*(pt(j)-xoff)**2
11    continue
      if (sum.lt.80.) then
        fxn=exp(-sum)
      else
        fxn=0.
      endif
      fxn=fxn*(5.64189**ndim)
      return
      END
```

Our demonstration of `miser`, the routine for multidimensional Monte Carlo integration with recursive stratified sampling, performs the same integral repeatedly, so as to test whether `miser`'s claimed standard deviation in fact corresponds (at least roughly) to the actual spread of its answers around the known correct value. The function being integrated is the same narrow Gaussian as was used for `vegas`. Now, however, we assume an offset x_0 that puts the peak of the Gaussian well into the region of integration, say $x_0 = 0.33333$.

The program prompts for the number of points to try in each integration, the number of dimensions, the value for x_0, a dithering parameter which can generally be set to zero (see discussion in main book), and the number of separate times to repeat the `miser` integration. A typical run with the input values

$$10000, \quad 4, \quad 0.333333, \quad 0.0, \quad 5$$

produces output like this:

```
FRACTIONAL ERROR: ACTUAL,INDICATED= 1.7880376E-02 2.1793477E-02
```

The first number is the average of the actual errors, comparing `miser`'s output value to the known answer of 1. The second number is the average of the standard deviations reported by `miser`, that is, its indicated error. In this example, we see that they are comparable, as expected, and that our Monte Carlo integration is being performed (each of five times) with an accuracy of about 2%.

Of course, in most real uses of `miser`, it would be better to perform the integration only once, with five times as many points, rather than five times.

```
      PROGRAM xmiser
C     driver for routine miser
      INTEGER idum,j,n,ndim,nt,ntries
      REAL ave,dith,sumav,sumsd,var,xoff
      COMMON /ranno/ idum
      COMMON /tofunc/ xoff,ndim
      REAL region(20)
      EXTERNAL func
      write(*,*) 'IDUM='
      read(*,*) idum
      idum=-abs(idum)
1     write(*,*) 'ENTER N,NDIM,XOFF,DITH,NTRIES'
      read(*,*,END=999) n,ndim,xoff,dith,ntries
      sumav=0.
      sumsd=0.
      do 12 nt=1,ntries
        do 11 j=1,ndim
          region(j)=0.
          region(j+ndim)=1.
11      continue
        call miser(func,region,ndim,n,dith,ave,var)
        sumav=sumav+(ave-1.)**2
        sumsd=sumsd+sqrt(abs(var))
12    continue
      sumav=sqrt(sumav/ntries)
      sumsd=sumsd/ntries
      write(*,*) 'FRACTIONAL ERROR: ACTUAL,INDICATED=',sumav,sumsd
      goto 1
999   write(*,*) 'NORMAL COMPLETION'
      STOP
      END

      REAL FUNCTION func(pt)
      INTEGER j,ndim
      REAL sum,xoff
      COMMON /tofunc/ xoff,ndim
      REAL pt(*)
      sum=0.
      do 11 j=1,ndim
```

```
         sum=sum+100.*(pt(j)-xoff)**2
11     continue
       if (sum.lt.80.) then
          func=exp(-sum)
       else
          func=0.
       endif
       func=func*(5.64189**ndim)
       return
       END
```

Chapter 8: Sorting

Chapter 8 of Numerical Recipes covers a variety of sorting tasks including sorting arrays into numerical order, preparing an index table for the order of an array, and preparing a rank table showing the rank order of each element in an array. piksrt *sorts a single array by the straight insertion method.* piksr2 *sorts by the same method but makes the corresponding rearrangement of a second array as well.* shell *carries out a Shell sort.* sort *and* sort2 *both do a quicksort, and they are related in the same way as* piksrt *and* piksr2, *that is,* sort *sorts a single array, and* sort2 *sorts an array while correspondingly rearranging a second array.* hpsort *sorts an array by the heapsort algorithm, which is not quite as fast (on average) as quicksort, but requires no auxiliary storage.*

indexx *indexes an array. That is, it produces a second array that contains pointers to the elements of the original array in the order of their size.* sort3 *uses* indexx *and illustrates its value by sorting one array while making corresponding rearrangements in two others.* rank *produces the rank table for an array of data. The rank table is a second array whose elements list the rank order of the corresponding elements of the original array.*

select *finds the Nth largest element in an array (e.g., the median). It uses the same partitioning idea as quicksort and is very fast.* selip *performs the same function without disturbing the array, and is correspondingly slower.* hpsel *finds the M largest elements of an array, without disturbing it.*

Finally, the routines eclass *and* eclazz *deal with equivalence classes.* eclass *gives the equivalence class of each element in an array based on a list of equivalent pairs that it is given as input.* eclazz *gives the same output but bases it on a procedure named* equiv(j,k) *that tells whether two array elements* j *and* k *are in the same equivalence class.*

$$\star \quad \star \quad \star \quad \star$$

Routine piksrt sorts an array by straight insertion. Sample program xpiksrt provides it with a 100-element array from file TARRAY.DAT which is listed in the Appendix to this chapter. The program prints both the original and the sorted array for comparison.

```
      PROGRAM xpiksrt
C     driver for routine piksrt
      INTEGER i,j
      REAL a(100)
      open(7,file='TARRAY.DAT',status='OLD')
      read(7,*) (a(i),i=1,100)
```

```
      close(7)
C     print original array
      write(*,*) 'Original array:'
      do 11 i=1,10
        write(*,'(1x,10f7.2)') (a(10*(i-1)+j),j=1,10)
11    continue
C     sort array
      call piksrt(100,a)
C     print sorted array
      write(*,*) 'Sorted array:'
      do 12 i=1,10
        write(*,'(1x,10f7.2)') (a(10*(i-1)+j),j=1,10)
12    continue
      END
```

piksr2 sorts an array, and simultaneously rearranges a second array (of the same size) correspondingly. In program xpiksr2, the first array a(i) is again taken from TARRAY.DAT. The second is defined by b(i)=i. In other words, b is originally sorted and a is not. After a call to piksr2, the situation should be reversed. With a second call, this time with b as the first argument and a as the second, the two arrays should be returned to their original form.

```
      PROGRAM xpiksr2
C     driver for routine piksr2
      INTEGER i,j
      REAL a(100),b(100)
      open(7,file='TARRAY.DAT',status='OLD')
      read(7,*) (a(i),i=1,100)
      close(7)
C     generate B-array
      do 11 i=1,100
        b(i)=i
11    continue
C     sort A and mix B
      call piksr2(100,a,b)
      write(*,*) 'After sorting A and mixing B, array A is:'
      do 12 i=1,10
        write(*,'(1x,10f7.2)') (a(10*(i-1)+j), j=1,10)
12    continue
      write(*,*) '...and array B is:'
      do 13 i=1,10
        write(*,'(1x,10f7.2)') (b(10*(i-1)+j), j=1,10)
13    continue
      write(*,*) 'press RETURN to continue...'
      read(*,*)
C     sort B and mix A
      call piksr2(100,b,a)
      write(*,*) 'After sorting B and mixing A, array A is:'
      do 14 i=1,10
        write(*,'(1x,10f7.2)') (a(10*(i-1)+j), j=1,10)
14    continue
      write(*,*) '...and array B is:'
      do 15 i=1,10
        write(*,'(1x,10f7.2)') (b(10*(i-1)+j), j=1,10)
15    continue
      END
```

Subroutine `shell` does a Shell sort of a data array. The calling format is identical to that of `piksrt`, and so we use the same sample program, now called `xshell`.

```
      PROGRAM xshell
C     driver for routine shell
      INTEGER i,j
      REAL a(100)
      open(7,file='TARRAY.DAT',status='OLD')
      read(7,*) (a(i),i=1,100)
      close(7)
C     print original array
      write(*,*) 'Original array:'
      do 11 i=1,10
        write(*,'(1x,10f7.2)') (a(10*(i-1)+j),j=1,10)
11    continue
C     sort array
      call shell(100,a)
C     print sorted array
      write(*,*) 'Sorted array:'
      do 12 i=1,10
        write(*,'(1x,10f7.2)') (a(10*(i-1)+j),j=1,10)
12    continue
      END
```

By the same token, routines `sort` and `sort2` employ the same programs as routines `piksrt` and `piksr2`, respectively. (Here they are called `xsort` and `xsort2`.) Both routines use the quicksort algorithm. `sort`, however, works on a single array. `sort2` sorts one array while making corresponding rearrangements to a second.

```
      PROGRAM xsort
C     driver for routine sort
      INTEGER i,j
      REAL a(100)
      open(7,file='TARRAY.DAT',status='OLD')
      read(7,*) (a(i),i=1,100)
      close(7)
C     print original array
      write(*,*) 'Original array:'
      do 11 i=1,10
        write(*,'(1x,10f7.2)') (a(10*(i-1)+j),j=1,10)
11    continue
C     sort array
      call sort(100,a)
C     print sorted array
      write(*,*) 'Sorted array:'
      do 12 i=1,10
        write(*,'(1x,10f7.2)') (a(10*(i-1)+j),j=1,10)
12    continue
      END

      PROGRAM xsort2
C     driver for routine sort2
      INTEGER i,j
      REAL a(100),b(100)
      open(7,file='TARRAY.DAT',status='OLD')
      read(7,*) (a(i),i=1,100)
      close(7)
```

```
C      generate B-array
       do 11 i=1,100
         b(i)=i
11     continue
C      sort A and mix B
       call sort2(100,a,b)
       write(*,*) 'After sorting A and mixing B, array A is:'
       do 12 i=1,10
         write(*,'(1x,10f7.2)') (a(10*(i-1)+j), j=1,10)
12     continue
       write(*,*) '...and array B is:'
       do 13 i=1,10
         write(*,'(1x,10f7.2)') (b(10*(i-1)+j), j=1,10)
13     continue
       write(*,*) 'press RETURN to continue...'
       read(*,*)
C      sort B and mix A
       call sort2(100,b,a)
       write(*,*) 'After sorting B and mixing A, array A is:'
       do 14 i=1,10
         write(*,'(1x,10f7.2)') (a(10*(i-1)+j), j=1,10)
14     continue
       write(*,*) '...and array B is:'
       do 15 i=1,10
         write(*,'(1x,10f7.2)') (b(10*(i-1)+j), j=1,10)
15     continue
       END
```

hpsort sorts an array by the heapsort algorithm. Its calling sequence is exactly like that of piksrt and sort, so we again rely on the same sample program, now called xhpsort.

```
       PROGRAM xhpsort
C      driver for routine hpsort
       INTEGER i,j
       REAL a(100)
       open(7,file='TARRAY.DAT',status='OLD')
       read(7,*) (a(i),i=1,100)
       close(7)
C      print original array
       write(*,*) 'Original array:'
       do 11 i=1,10
         write(*,'(1x,10f7.2)') (a(10*(i-1)+j),j=1,10)
11     continue
C      sort array
       call hpsort(100,a)
C      print sorted array
       write(*,*) 'Sorted array:'
       do 12 i=1,10
         write(*,'(1x,10f7.2)') (a(10*(i-1)+j),j=1,10)
12     continue
       END
```

The subroutine indexx generates the index array for a given input array. The index array indx(j) gives, for each j, the index of the element of the input array which will assume position j if the array is sorted. That is, for an input array a, the sorted version of a will be a(indx(j)). To demonstrate this, sample program xindexx produces an index for the array in TARRAY.DAT. It then prints the array in the order a(indx(j)), j=1,..,100 for inspection.

```
      PROGRAM xindexx
C     driver for routine indexx
      INTEGER i,j,indx(100)
      REAL a(100)
      open(7,file='TARRAY.DAT',status='OLD')
      read(7,*) (a(i),i=1,100)
      close(7)
C     generate index for sorted array
      call indexx(100,a,indx)
C     print original array
      write(*,*) 'Original array:'
      do 11 i=1,10
        write(*,'(1x,10f7.2)') (a(10*(i-1)+j),j=1,10)
11    continue
C     print sorted array
      write(*,*) 'Sorted array:'
      do 12 i=1,10
        write(*,'(1x,10f7.2)') (a(indx(10*(i-1)+j)),j=1,10)
12    continue
      END
```

One use for indexx is the management of more than two arrays. sort3, for example, sorts one array while making corresponding reorderings of two other arrays. In sample program xsort3, the first array is taken as the first 64 elements of TARRAY.DAT (see Appendix). The second and third arrays are taken to be the numbers 1 to 64 in forward order and reverse order, respectively. When the first array is ordered, the second and third are scrambled, but scrambled in exactly the same way. To prove this, a text message is assigned to a character array. Then the letters are scrambled according to the order of numbers found in the rearranged second array. They are subsequently unscrambled according to the order of numbers found in the rearranged third array. If sort3 works properly, this ought to leave the message reading in the reverse order.

```
      PROGRAM xsort3
C     driver for routine sort3
      INTEGER NLEN
      PARAMETER(NLEN=64)
      INTEGER i,j,indx(NLEN)
      REAL a(NLEN),b(NLEN),c(NLEN),wksp(NLEN)
      CHARACTER msg1*33,msg2*31
      CHARACTER msg*64,amsg(64)*1,bmsg(64)*1,cmsg(64)*1
      EQUIVALENCE(msg,amsg(1)),(msg1,amsg(1)),(msg2,amsg(34))
      DATA msg1/'I''d rather have a bottle in front'/
      DATA msg2/' of me than a frontal lobotomy.'/
      write(*,*) 'Original message:'
      write(*,'(1x,64a1,/)') (amsg(j),j=1,64)
C     read array of random numbers
      open(7,file='TARRAY.DAT',status='OLD')
      read(7,*) (a(i),i=1,NLEN)
      close(7)
C     create array B and array C
      do 11 i=1,NLEN
```

```
          b(i)=i
          c(i)=NLEN+1-i
11      continue
C       sort array A while mixing B and C
        call sort3(NLEN,a,b,c,wksp,indx)
C       scramble message according to array B
        do 12 i=1,NLEN
          j=b(i)
          bmsg(i)=amsg(j)
12      continue
        write(*,*) 'Scrambled message:'
        write(*,'(1x,64a1,/)') (bmsg(j),j=1,64)
C       unscramble according to array C
        do 13 i=1,NLEN
          j=c(i)
          cmsg(j)=bmsg(i)
13      continue
        write(*,*) 'Mirrored message:'
        write(*,'(1x,64a1,/)') (cmsg(j),j=1,64)
        END
```

rank is a subroutine that is similar to indexx. Instead of producing an indexing array, though, it produces a rank table. For an array a(j) and rank table irank(j), entry j in irank will tell what index a(j) will have if a is sorted. irank actually takes its input information not from the array itself, but from the index array produced by indexx. Sample program xrank begins with the array from TARRAY.DAT, and feeds it to indexx and rank. The table of ranks produced is listed. To check it, the array a is copied into an array b in the rank order suggested by irank. b should then be in proper order.

```
        PROGRAM xrank
C       driver for routine rank
        INTEGER i,j,k,l,indx(100),irank(100)
        REAL a(100),b(10)
        open(7,file='TARRAY.DAT',status='OLD')
        read(7,*) (a(i),i=1,100)
        close(7)
        call indexx(100,a,indx)
        call rank(100,indx,irank)
        write(*,*) 'Original array is:'
        do 11 i=1,10
          write(*,'(1x,10f7.2)') (a(10*(i-1)+j), j=1,10)
11      continue
        write(*,*) 'Table of ranks is:'
        do 12 i=1,10
          write(*,'(1x,10i6)') (irank(10*(i-1)+j), j=1,10)
12      continue
        write(*,*) 'press RETURN to continue...'
        read(*,*)
        write(*,*) 'Array sorted according to rank table:'
        do 15 i=1,10
          do 14 j=1,10
            k=10*(i-1)+j
            do 13 l=1,100
              if (irank(l).eq.k) b(j)=a(l)
13          continue
```

```
14      continue
        write(*,'(1x,10f7.2)') (b(j),j=1,10)
15      continue
        END
```

Sample program `xselect` demonstrates the use of `select` to find the kth largest element in `TARRAY.DAT`. As a cross-check, the answer is compared with that from the slower routine `selip`.

```
        PROGRAM xselect
C       driver for routine select
        INTEGER NP
        PARAMETER(NP=100)
        INTEGER i,k
        REAL q,s,select,selip,arr(NP),brr(NP)
        open(7,file='TARRAY.DAT',status='OLD')
        read(7,*) (arr(i),i=1,NP)
        close(7)
1       write(*,*) 'INPUT K'
        read(*,*,END=999) k
        do 11 i=1,NP
          brr(i)=arr(i)
11      continue
        s=selip(k,NP,brr)
        q=select(k,NP,brr)
        write(*,*) 'Element in sort position',k,' is',q
        write(*,*) 'Cross-check from SELIP routine',s
        goto 1
999     write(*,*) 'NORMAL COMPLETION'
        STOP
        END
```

The routine `xselip` tests `selip` by selecting the kth largest element of TARRAY.DAT, for each $k = 1, \ldots, 100$. This is a very inefficient way of sorting the array. As a final check, `xselip` prints out a user-supplied request for the kth largest element of the array.

```
        PROGRAM xselip
C       driver for routine selip
        INTEGER i,j,k
        REAL q,selip,arr(100),brr(100)
        open(7,file='TARRAY.DAT',status='OLD')
        read(7,*) (arr(i),i=1,100)
        close(7)
C       print original array
        write(*,*) 'Original array:'
        do 11 i=1,10
          write(*,'(1x,10f7.2)') (arr(10*(i-1)+j),j=1,10)
11      continue
C       sort array - inefficiently, but shows use and verifies routine
        do 12 i=1,100
          brr(i)=selip(i,100,arr)
12      continue
C       print sorted array
        write(*,*) 'Sorted array:'
        do 13 i=1,10
          write(*,'(1x,10f7.2)') (brr(10*(i-1)+j),j=1,10)
```

```
13      continue
1       write(*,*) 'INPUT K'
        read(*,*,END=999) k
        q=selip(k,100,arr)
        write(*,*) 'Element in sort position',k,' is',q
        goto 1
999     write(*,*) 'NORMAL COMPLETION'
        STOP
        END
```

Routine hpsel finds the k largest elements of an array, where (for efficiency) k should be much shorter than the length of the array. The sample program is much like that for select, except that all k largest elements (not just the kth) are printed out.

```
        PROGRAM xhpsel
C       driver for routine hpsel
        INTEGER NP
        PARAMETER(NP=100)
        REAL arr(NP),heap(NP)
        INTEGER i,k
        REAL check,select
        open(7,file='TARRAY.DAT',status='OLD')
        read(7,*) (arr(i),i=1,NP)
        close(7)
1       write(*,*) 'enter K'
        read(*,*,END=999) k
        call hpsel(k,NP,arr,heap)
        check=select(NP+1-k,NP,arr)
        write(*,*) 'heap(1),check=',heap(1),check
        write(*,*) 'heap of numbers of size',k
        do 11 i=1,k
          write(*,*) i,heap(i)
11      continue
        goto 1
999     write(*,*) 'NORMAL COMPLETION'
        STOP
        END
```

Subroutine eclass generates a list of equivalence classes for the elements of an input array, based on the arrays lista(j) and listb(j) which list equivalent pairs for each j. In sample program xeclass, these lists are

$$\text{lista}: \quad 1, 1, 5, 2, 6, 2, 7, 11, 3, 4, 12$$
$$\text{listb}: \quad 5, 9, 13, 6, 10, 14, 3, 7, 15, 8, 4$$

According to these lists, 1 is equivalent to 5, 1 is equivalent to 9, etc. If you work it out, you will find the following classes:

$$\text{class1}: \quad 1, 5, 9, 13$$
$$\text{class2}: \quad 2, 6, 10, 14$$
$$\text{class3}: \quad 3, 7, 11, 15$$
$$\text{class4}: \quad 4, 8, 12$$

The sample program prints out the classes and ought to agree with this list.

```
      PROGRAM xeclass
C     driver for routine eclass
      INTEGER N,M
      PARAMETER(N=15,M=11)
      INTEGER i,j,k,lclas,nclass,
     *    lista(M),listb(M),nf(N),nflag(N),nsav(N)
      DATA lista/1,1,5,2,6,2,7,11,3,4,12/
      DATA listb/5,9,13,6,10,14,3,7,15,8,4/
      call eclass(nf,N,lista,listb,M)
      do 11 i=1,N
        nflag(i)=1
11    continue
      write(*,'(/1x,a)') 'Numbers from 1-15 divided according to'
      write(*,'(1x,a/)') 'their value modulo 4:'
      lclas=0
      do 13 i=1,N
        nclass=nf(i)
        if (nflag(nclass).ne.0) then
          nflag(nclass)=0
          lclas=lclas+1
          k=0
          do 12 j=i,N
            if (nf(j).eq.nf(i)) then
              k=k+1
              nsav(k)=j
            endif
12        continue
          write(*,'(1x,a,i2,a,3x,5i3)') 'Class',
     *          lclas,':',(nsav(j),j=1,k)
        endif
13    continue
      END
```

eclazz performs the same analysis but figures the equivalences from a logical function equiv(i,j) that tells whether i and j are in the same equivalence class. In xeclazz, equiv is defined as .true. if (i mod 4) and (j mod 4) are the same. It is otherwise .false.

```
      PROGRAM xeclazz
C     driver for routine eclazz
      LOGICAL equiv
      INTEGER N
      PARAMETER(N=15)
      INTEGER i,j,k,lclas,nclass,nf(N),nflag(N),nsav(N)
      EXTERNAL equiv
      call eclazz(nf,N,equiv)
      do 11 i=1,N
        nflag(i)=1
11    continue
      write(*,'(/1x,a)') 'Numbers from 1-15 divided according to'
      write(*,'(1x,a/)') 'their value modulo 4:'
      lclas=0
      do 13 i=1,N
        nclass=nf(i)
        if (nflag(nclass).ne.0) then
          nflag(nclass)=0
          lclas=lclas+1
```

```
        k=0
        do 12 j=i,N
          if (nf(j).eq.nf(i)) then
            k=k+1
            nsav(k)=j
          endif
12      continue
        write(*,'(1x,a,i2,a,3x,5i3)') 'Class',
     *          lclas,':',(nsav(j),j=1,k)
      endif
13    continue
      END

      LOGICAL FUNCTION equiv(i,j)
      INTEGER i,j
      equiv=.false.
      if (mod(i,4).eq.mod(j,4)) equiv=.true.
      return
      END
```

Appendix

File TARRAY.DAT:

```
29.82 71.51  3.30 87.44 53.42 63.16 89.10 25.75 93.16 27.72
71.58 48.34 53.11 18.34 27.13 60.31 83.34 22.81 66.84 52.91
53.42 15.22  8.01 53.39 76.12 79.09 67.61 38.39 24.81 73.21
13.42 52.10 34.86 99.83 38.46 81.59 61.75 79.62 93.39  3.21
99.34 92.22 94.29  7.03  6.67 89.35 83.14  9.01 12.68 62.22
 2.95 85.02 95.82 73.96 49.29 77.72 36.65  3.48 48.98 71.83
 1.41  9.48 32.37 89.95 28.39 79.36 54.05 46.08 11.67 37.78
77.17 74.33 10.13  4.62 49.95 68.40 19.40 34.06  4.11 98.40
42.44 64.14 89.41 52.99 71.79  3.94 19.73 44.91 71.44 59.10
27.54 15.67 67.95 55.61 26.05 25.01 82.09 89.67 57.08 38.27
```

Chapter 9: Root Finding and Sets of Equations

Chapter 9 of Numerical Recipes deals primarily with the problem of finding roots to equations, and treats the problem in greatest detail in one dimension. We begin with a general-purpose routine called scrsho *that produces a crude graph of a given function on a specified interval. It is used for low-resolution plotting to investigate the properties of the function. With this in hand, we add bracketing routines* zbrac *and* zbrak. *The first of these takes a function and an interval, and expands the interval geometrically until it brackets a root. The second breaks the interval into N subintervals of equal size. It then reports any intervals that contain at least one root. Once bracketed, roots can be found by a number of other routines.* rtbis *finds such roots by bisection.* rtflsp *and* rtsec *use the method of false position and the secant method, respectively.* zriddr *implements Ridders' method.* zbrent *uses a combination of methods to give assured and relatively efficient convergence.* rtnewt *implements the Newton-Raphson root finding method, while* rtsafe *combines it with bisection to correct for its risky global convergence properties.*

For finding the roots of polynomials, laguer *is handy, and when combined with its driver* zroots *it can find all roots of a polynomial having complex coefficients. An alternative to* laguer *is provided by* zrhqr, *which finds the roots as eigenvalues of the companion matrix. When you have some tentative complex roots of a real polynomial, they can be polished by* qroot, *which employs Bairstow's method.*

In multiple dimensions, root-finding requires some foresight. However, if you can identify the neighborhood of a root of a system of nonlinear equations, then mnewt *will help you to zero in using Newton-Raphson. More powerful is to use the globally convergent multi-dimensional Newton method in* newt, *or Broyden's method* broydn, *which generalizes the secant method.*

$$\star \quad \star \quad \star \quad \star$$

scrsho is a primitive graphing routine that will print graphs on virtually any terminal or printer. Sample program xscrsho demonstrates it by graphing the zero-order Bessel function J_0.

```
       PROGRAM xscrsho
C      driver for routine scrsho
       REAL bessj0
       EXTERNAL bessj0
       write(*,*) 'Graph of the Bessel Function J0:'
       call scrsho(bessj0)
```

```
      END
```

zbrac is a root-bracketing routine that works by expanding the range of an interval geometrically until it brackets a root. Sample program xzbrac applies it to the Bessel function J_0. It starts with the ten intervals (1.0, 2.0), (2.0, 3.0), etc., and expands each until it contains a root. Then it prints the interval limits, and the function J_0 evaluated at these limits. The two values of J_0 should have opposite signs.

```
      PROGRAM xzbrac
C     driver for routine zbrac
      LOGICAL succes
      INTEGER i
      REAL bessj0,x1,x2
      EXTERNAL bessj0
      write(*,'(/1x,t4,a,t29,a/)') 'Bracketing values:',
     *     'Function values:'
      write(*,'(1x,t6,a,t16,a,t29,a,t41,a/)') 'X1','X2',
     *     'BESSJ0(X1)','BESSJ0(X2)'
      do 11 i=1,10
        x1=i
        x2=x1+1.0
        call zbrac(bessj0,x1,x2,succes)
        if (succes) then
          write(*,'(1x,f7.2,f10.2,7x,2f12.6)') x1,x2,
     *            bessj0(x1),bessj0(x2)
        endif
11    continue
      END
```

zbrak is much like zbrac except that it takes an interval and subdivides it into N equal parts. It then identifies any of the subintervals that contain roots. Sample program xzbrak looks for roots of $J_0(x)$ between X1 = 1.0 and X2 = 50.0 by allowing zbrak to divide the interval into $N = 100$ parts. If there are no roots spaced closer than $\Delta x = 0.49$, then it will find brackets for all roots in this region. The limits of bracketing intervals, as well as function values at these limits, are printed, and again the function values at the end of each interval ought to be of opposite sign. There are 16 roots of J_0 between 1 and 50.

```
      PROGRAM xzbrak
C     driver for routine zbrak
      INTEGER N,NBMAX
      REAL X1,X2
      PARAMETER(N=100,NBMAX=20,X1=1.0,X2=50.0)
      INTEGER i,nb
      REAL bessj0,xb1(NBMAX),xb2(NBMAX)
      EXTERNAL bessj0
      nb=NBMAX
      call zbrak(bessj0,X1,X2,N,xb1,xb2,nb)
      write(*,'(/1x,a/)') 'Brackets for roots of BESSJ0:'
      write(*,'(/1x,t17,a,t27,a,t40,a,t50,a/)') 'lower','upper',
     *     'F(lower)','F(upper)'
      do 11 i=1,nb
        write(*,'(1x,a,i2,2(4x,2f10.4))') 'Root ',i,xb1(i),xb2(i),
     *          bessj0(xb1(i)),bessj0(xb2(i))
11    continue
```

```
      END
```

Routine `rtbis` begins with the brackets for a root and finds the root itself by bisection, The accuracy with which the root is found is determined by parameter `xacc`. Sample program `xrtbis` finds all the roots of Bessel function $J_0(x)$ between $X1 = 1.0$ and $X2 = 50.0$. In this case `xacc` is specified to be about 10^{-6} of the value of the root itself (actually, 10^{-6} of the center of the interval being bisected). The roots `root` are listed, as well as $J_0(\text{root})$ to verify their accuracy.

```
      PROGRAM xrtbis
C     driver for routine rtbis
      INTEGER N,NBMAX
      REAL X1,X2
      PARAMETER(N=100,NBMAX=20,X1=1.0,X2=50.0)
      INTEGER i,nb
      REAL bessj0,rtbis,root,xacc,xb1(NBMAX),xb2(NBMAX)
      EXTERNAL bessj0
      nb=NBMAX
      call zbrak(bessj0,X1,X2,N,xb1,xb2,nb)
      write(*,'(/1x,a)') 'Roots of BESSJ0:'
      write(*,'(/1x,t19,a,t31,a/)') 'x','F(x)'
      do 11 i=1,nb
        xacc=(1.0e-6)*(xb1(i)+xb2(i))/2.0
        root=rtbis(bessj0,xb1(i),xb2(i),xacc)
        write(*,'(1x,a,i2,2x,f12.6,e16.4)') 'Root ',i,root,bessj0(root)
11    continue
      END
```

The next six sample programs are essentially identical to the one just discussed, except for the root-finder they employ. `xrtflsp` calls `rtflsp`, finding the root by "false position". `xrtsec` calls `rtsec` and uses the secant method. `xzriddr` uses Ridders' method, `zriddr`. `xzbrent` uses `zbrent` to give reliable and efficient convergence. The Newton-Raphson method implemented in `rtnewt` is demonstrated by `xrtnewt`, and `xrtsafe` calls `rtsafe`, which improves upon `rtnewt` by combining it with bisection to achieve better global convergence. The latter two programs include a subroutine `funcd` that returns the value of the function and its derivative at a given x. In the case of test function $J_0(x)$ the derivative is $-J_1(x)$, and is conveniently in our collection of special functions.

```
      PROGRAM xrtflsp
C     driver for routine rtflsp
      INTEGER N,NBMAX
      REAL X1,X2
      PARAMETER(N=100,NBMAX=20,X1=1.0,X2=50.0)
      INTEGER i,nb
      REAL bessj0,rtflsp,root,xacc,xb1(NBMAX),xb2(NBMAX)
      EXTERNAL bessj0
      nb=NBMAX
      call zbrak(bessj0,X1,X2,N,xb1,xb2,nb)
      write(*,'(/1x,a)') 'Roots of BESSJ0:'
      write(*,'(/1x,t19,a,t31,a/)') 'x','F(x)'
      do 11 i=1,nb
        xacc=(1.0e-6)*(xb1(i)+xb2(i))/2.0
        root=rtflsp(bessj0,xb1(i),xb2(i),xacc)
        write(*,'(1x,a,i2,2x,f12.6,e16.4)') 'Root ',i,root,bessj0(root)
```

```
11      continue
        END


        PROGRAM xrtsec
C       driver for routine rtsec
        INTEGER N,NBMAX
        REAL X1,X2
        PARAMETER(N=100,NBMAX=20,X1=1.0,X2=50.0)
        INTEGER i,nb
        REAL bessj0,rtsec,root,xacc,xb1(NBMAX),xb2(NBMAX)
        EXTERNAL bessj0
        nb=NBMAX
        call zbrak(bessj0,X1,X2,N,xb1,xb2,nb)
        write(*,'(/1x,a)') 'Roots of BESSJ0:'
        write(*,'(/1x,t19,a,t31,a/)') 'x','F(x)'
        do 11 i=1,nb
          xacc=(1.0e-6)*(xb1(i)+xb2(i))/2.0
          root=rtsec(bessj0,xb1(i),xb2(i),xacc)
          write(*,'(1x,a,i2,2x,f12.6,e16.4)') 'Root ',i,root,bessj0(root)
11      continue
        END


        PROGRAM xzriddr
C       driver for routine zriddr
        INTEGER N,NBMAX
        REAL X1,X2
        PARAMETER(N=100,NBMAX=20,X1=1.0,X2=50.0)
        INTEGER i,nb
        REAL bessj0,zriddr,root,xacc,xb1(NBMAX),xb2(NBMAX)
        EXTERNAL bessj0
        nb=NBMAX
        call zbrak(bessj0,X1,X2,N,xb1,xb2,nb)
        write(*,'(/1x,a)') 'Roots of BESSJ0:'
        write(*,'(/1x,t19,a,t31,a/)') 'x','F(x)'
        do 11 i=1,nb
          xacc=(1.0e-6)*(xb1(i)+xb2(i))/2.0
          root=zriddr(bessj0,xb1(i),xb2(i),xacc)
          write(*,'(1x,a,i2,2x,f12.6,e16.4)') 'Root ',i,root,bessj0(root)
11      continue
        END


        PROGRAM xzbrent
C       driver for routine zbrent
        INTEGER N,NBMAX
        REAL X1,X2
        PARAMETER(N=100,NBMAX=20,X1=1.0,X2=50.0)
        INTEGER i,nb
        REAL bessj0,zbrent,root,tol,xb1(NBMAX),xb2(NBMAX)
        EXTERNAL bessj0
        nb=NBMAX
        call zbrak(bessj0,X1,X2,N,xb1,xb2,nb)
        write(*,'(/1x,a)') 'Roots of BESSJ0:'
        write(*,'(/1x,t19,a,t31,a/)') 'x','F(x)'
        do 11 i=1,nb
          tol=(1.0e-6)*(xb1(i)+xb2(i))/2.0
          root=zbrent(bessj0,xb1(i),xb2(i),tol)
          write(*,'(1x,a,i2,2x,f12.6,e16.4)') 'Root ',i,root,bessj0(root)
```

```
11      continue
        END


        PROGRAM xrtnewt
C       driver for routine rtnewt
        INTEGER N,NBMAX
        REAL X1,X2
        PARAMETER(N=100,NBMAX=20,X1=1.0,X2=50.0)
        INTEGER i,nb
        REAL bessj0,rtnewt,root,xacc,xb1(NBMAX),xb2(NBMAX)
        EXTERNAL funcd,bessj0
        nb=NBMAX
        call zbrak(bessj0,X1,X2,N,xb1,xb2,nb)
        write(*,'(/1x,a)') 'Roots of BESSJ0:'
        write(*,'(/1x,t19,a,t31,a/)') 'x','F(x)'
        do 11 i=1,nb
          xacc=(1.0e-6)*(xb1(i)+xb2(i))/2.0
          root=rtnewt(funcd,xb1(i),xb2(i),xacc)
          write(*,'(1x,a,i2,2x,f12.6,e16.4)') 'Root ',i,root,bessj0(root)
11      continue
        END


        SUBROUTINE funcd(x,fn,df)
        REAL bessj0,bessj1,df,fn,x
        fn=bessj0(x)
        df=-bessj1(x)
        return
        END


        PROGRAM xrtsafe
C       driver for routine rtsafe
        INTEGER N,NBMAX
        REAL X1,X2
        PARAMETER(N=100,NBMAX=20,X1=1.0,X2=50.0)
        INTEGER i,nb
        REAL bessj0,rtsafe,root,xacc,xb1(NBMAX),xb2(NBMAX)
        EXTERNAL funcd,bessj0
        nb=NBMAX
        call zbrak(bessj0,X1,X2,N,xb1,xb2,nb)
        write(*,'(/1x,a)') 'Roots of BESSJ0:'
        write(*,'(/1x,t19,a,t31,a/)') 'x','F(x)'
        do 11 i=1,nb
          xacc=(1.0e-6)*(xb1(i)+xb2(i))/2.0
          root=rtsafe(funcd,xb1(i),xb2(i),xacc)
          write(*,'(1x,a,i2,2x,f12.6,e16.4)') 'Root ',i,root,bessj0(root)
11      continue
        END


        SUBROUTINE funcd(x,fn,df)
        REAL bessj0,bessj1,df,fn,x
        fn=bessj0(x)
        df=-bessj1(x)
        return
        END
```

Routine `laguer` finds the roots of a polynomial with complex coefficients. The polynomial of degree M is specified by $M + 1$ coefficients which, in sample program xlaguer, are specified in a DATA statement and kept in array a. The polynomial in this case is

$$F(x) = x^4 - (1 + 2i)x^2 + 2i$$

The four roots of this polynomial are $x = 1.0$, $x = -1.0$, $x = 1 + i$, and $x = -(1 + i)$. laguer proceeds on the basis of a trial root, and attempts to converge to true roots. The root to which it converges depends on the trial value. The program tries a series of complex trial values along the line in the imaginary plane from $-1.0 - i$ to $1.0 + i$. The actual roots to which it converges are compared to all previously found values, and if different, are printed.

```
      PROGRAM xlaguer
C     driver for routine laguer
      INTEGER M,MP1,NTRY
      REAL EPS
      PARAMETER(M=4,MP1=M+1,NTRY=21,EPS=1.0E-6)
      INTEGER i,iflag,its,j,n
      COMPLEX a(MP1),y(NTRY),x
      DATA a/(0.0,2.0),(0.0,0.0),(-1.0,-2.0),(0.0,0.0),(1.0,0.0)/
      write(*,'(/1x,a)') 'Roots of polynomial x^4-(1+2i)*x^2+2i'
      write(*,'(/1x,t16,a4,t29,a7,t39,a5/)') 'Real','Complex','#iter'
      n=0
      do 12 i=1,NTRY
        x=cmplx((i-11.0)/10.0,(i-11.0)/10.0)
        call laguer(a,M,x,its)
        if (n.eq.0) then
          n=1
          y(1)=x
          write(*,'(1x,i5,2f15.6,i5)') n,x,its
        else
          iflag=0
          do 11 j=1,n
            if (abs(x-y(j)).le.EPS*abs(x)) iflag=1
11        continue
          if (iflag.eq.0) then
            n=n+1
            y(n)=x
            write(*,'(1x,i5,2f15.6,i5)') n,x,its
          endif
        endif
12    continue
      END
```

zroots is a driver for laguer. Sample program xzroots exercises zroots, using the same polynomial as the previous routine. First it finds the four roots. Then it corrupts each one by multiplying by $1 + .01i$. Finally it uses zroots again to polish the corrupted roots by setting the logical parameter polish to .true.

```
      PROGRAM xzroots
C     driver for routine zroots
      INTEGER M,M1
      PARAMETER(M=4,M1=M+1)
      INTEGER i
      COMPLEX a(M1),roots(M)
      LOGICAL polish
      DATA a/(0.0,2.0),(0.0,0.0),(-1.0,-2.0),(0.0,0.0),(1.0,0.0)/
      write(*,'(/1x,a)') 'Roots of the polynomial x^4-(1+2i)*x^2+2i'
```

```
      polish=.false.
      call zroots(a,M,roots,polish)
      write(*,'(/1x,a)') 'Unpolished roots:'
      write(*,'(1x,t10,a,t25,a,t37,a)') 'Root #','Real','Imag.'
      do 11 i=1,M
        write(*,'(1x,i11,5x,2f12.6)') i,roots(i)
11    continue
      write(*,'(/1x,a)') 'Corrupted roots:'
      do 12 i=1,M
        roots(i)=roots(i)*(1.0+0.01*i)
12    continue
      write(*,'(1x,t10,a,t25,a,t37,a)') 'Root #','Real','Imag.'
      do 13 i=1,M
        write(*,'(1x,i11,5x,2f12.6)') i,roots(i)
13    continue
      polish=.true.
      call zroots(a,M,roots,polish)
      write(*,'(/1x,a)') 'Polished roots:'
      write(*,'(1x,t10,a,t25,a,t37,a)') 'Root #','Real','Imag.'
      do 14 i=1,M
        write(*,'(1x,i11,5x,2f12.6)') i,roots(i)
14    continue
      END
```

Routine zrhqr is an alternative to laguer for polynomials with real coefficients. Sample program xzrhqr tries it out on the polynomial $x^4 - 1$.

```
      PROGRAM xzrhqr
C     driver for routine zrhqr
      INTEGER M,MP1,NTRY
      PARAMETER(M=4,MP1=M+1,NTRY=21)
      INTEGER i
      REAL a(MP1),rtr(M),rti(M)
      DATA a/-1.0,0.0,0.0,0.0,1.0/
      write(*,'(/1x,a)') 'Roots of polynomial x^4-1'
      write(*,'(/1x,t16,a,t29,a/)') 'Real','Complex'
      call zrhqr(a,M,rtr,rti)
      do 11 i=1,M
        write(*,'(1x,i5,2f15.6)') i,rtr(i),rti(i)
11    continue
      END
```

qroot is used for finding quadratic factors of polynomials with real coefficients. In the case of sample program xqroot, the polynomial is

$$P(x) = x^6 - 6x^5 + 16x^4 - 24x^3 + 25x^2 - 18x + 10.$$

The program proceeds like that of laguer. Successive trial values for quadratic factors $x^2 + Bx + C$ (in the form of guesses for B and C) are made, and for each trial, qroot converges on correct values. If the B and C which are found are unlike any previous values, then they are printed. By this means, all three quadratic factors are located. The answer should be

$$P(x) = (x^2 + 1)(x^2 - 4x + 5)(x^2 - 2x + 2)$$

```
      PROGRAM xqroot
C     driver for routine qroot
      INTEGER N,NTRY
      REAL EPS,TINY
      PARAMETER(N=7,EPS=1.0e-6,NTRY=10,TINY=1.0e-5)
      INTEGER i,j,nflag,nroot
      REAL p(N),b(NTRY),c(NTRY)
      DATA p/10.0,-18.0,25.0,-24.0,16.0,-6.0,1.0/
      write(*,'(/1x,a)') 'P(x)=x^6-6x^5+16x^4-24x^3+25x^2-18x+10'
      write(*,'(1x,a)') 'Quadratic factors x^2+Bx+C'
      write(*,'(/1x,a,t15,a,t27,a/)') 'Factor','B','C'
      nroot=0
      do 12 i=1,NTRY
        c(i)=0.5*i
        b(i)=-0.5*i
        call qroot(p,N,b(i),c(i),EPS)
        if (nroot.eq.0) then
          write(*,'(1x,i3,2x,2f12.6)') nroot,b(i),c(i)
          nroot=1
        else
          nflag=0
          do 11 j=1,nroot
            if (abs(b(i)-b(j)).lt.TINY.and.abs(c(i)-c(j)).lt.TINY)
     *          nflag=1
11        continue
          if (nflag.eq.0) then
            write(*,'(1x,i3,2x,2f12.6)') nroot,b(i),c(i)
            nroot=nroot+1
          endif
        endif
12    continue
      END
```

mnewt looks for roots of multiple nonlinear equations. In order to run the sample program xmnewt we supply a subroutine usrfun that returns the matrix alpha of partial derivatives of the functions with respect to each of the variables, and vector beta, containing the negatives of the function values. The sample program tries to find sets of variables that solve the four equations

$$-x_1^2 - x_2^2 - x_3^2 + x_4 = 0$$
$$x_1^2 + x_2^2 + x_3^2 + x_4^2 - 1 = 0$$
$$x_1 - x_2 = 0$$
$$x_2 - x_3 = 0$$

You will probably be able to find the two solutions to this set even without mnewt, noting that $x_1 = x_2$ and $x_2 = x_3$. If not, simply take the output from mnewt and plug it into these equations for verification. The output from mnewt should convince you of the need for good starting values (or that you should use newt instead!).

```
      PROGRAM xmnewt
C     driver for routine mnewt
      INTEGER NTRIAL,N,NP
      REAL TOLX,TOLF
      PARAMETER(NTRIAL=5,TOLX=1.0E-6,N=4,TOLF=1.0E-6,NP=15)
      INTEGER i,j,k,kk
```

```
      REAL xx,fjac(NP,NP),fvec(NP),x(NP)
      do 15 kk=-1,1,2
        do 14 k=1,3
          xx=0.2001*k*kk
          write(*,'(/1x,a,i2)') 'Starting vector number',k
          do 11 i=1,N
            x(i)=xx+0.2*i
            write(*,'(1x,t5,a,i1,a,f5.2)') 'X(',i,') = ',x(i)
11        continue
          do 13 j=1,NTRIAL
            call mnewt(1,x,N,TOLX,TOLF)
            call usrfun(x,n,NP,fvec,fjac)
            write(*,'(/1x,t5,a,t14,a,t29,a/)') 'I','X(I)','F'
            do 12 i=1,N
              write(*,'(1x,i4,2e15.6)') i,x(i),fvec(i)
12          continue
            write(*,'(/1x,a)') 'press RETURN to continue...'
            read(*,*)
13        continue
14      continue
15    continue
      END

      SUBROUTINE usrfun(x,n,np,fvec,fjac)
      INTEGER i,n,np
      REAL fjac(np,np),fvec(np),x(np)
      fjac(1,1)=-2.0*x(1)
      fjac(1,2)=-2.0*x(2)
      fjac(1,3)=-2.0*x(3)
      fjac(1,4)=1.0
      do 11 i=1,n
        fjac(2,i)=2.0*x(i)
11    continue
      fjac(3,1)=1.0
      fjac(3,2)=-1.0
      fjac(3,3)=0.0
      fjac(3,4)=0.0
      fjac(4,1)=0.0
      fjac(4,2)=1.0
      fjac(4,3)=-1.0
      fjac(4,4)=0.0
      fvec(1)=-x(1)**2-x(2)**2-x(3)**2+x(4)
      fvec(2)=x(1)**2+x(2)**2+x(3)**2+x(4)**2-1.0
      fvec(3)=x(1)-x(2)
      fvec(4)=x(2)-x(3)
      END
```

Sample program `xnewt` tries out routine `newt` on the nonlinear system

$$f_1 = x_1^2 + x_2^2 - 2$$
$$f_2 = e^{(x_1-1)} + x_2^3 - 2$$

which has a root at $(x_1, x_2) = (1, 1)$. It attempts to find the root starting from the initial guess $(2, .5)$. (This is Example 6.5.1 in *Numerical Methods for Unconstrained Optimization and Nonlinear Equations* by Dennis and Schnabel.) `mnewt`, which applies Newton's method blindly without a globally convergent strategy, fails on this example. (Note that `newt` is also used in the solution of two point boundary value problems in Chapter 17.)

```
      PROGRAM xnewt
C     driver for routine newt
      INTEGER N
      PARAMETER(N=2)
      INTEGER i
      REAL x(N),f(N)
      LOGICAL check
      x(1)=2.
      x(2)=.5
      call newt(x,N,check)
      call funcv(N,x,f)
      if (check) then
        write(*,*) 'Convergence problems.'
      endif
      write(*,'(1x,a5,t10,a1,t22,a1)') 'Index','x','f'
      do 11 i=1,N
        write(*,'(1x,i2,2x,2f12.6)') i,x(i),f(i)
11    continue
      END

      SUBROUTINE funcv(n,x,f)
      INTEGER n
      REAL x(n),f(n)
      f(1)=x(1)**2+x(2)**2-2.
      f(2)=exp(x(1)-1.)+x(2)**3-2.
      return
      END
```

The sample program for `broydn`, which implements a multidimensional generalization of the secant method, is identical to the one for `newt`:

```
      PROGRAM xbroydn
C     driver for routine broydn
      INTEGER N
      PARAMETER(N=2)
      INTEGER i
      REAL x(N),f(N)
      LOGICAL check
      x(1)=2.
      x(2)=.5
      call broydn(x,N,check)
      call funcv(N,x,f)
      if (check) then
        write(*,*) 'Convergence problems.'
      endif
      write(*,'(1x,a5,t10,a1,t22,a1)') 'Index','x','f'
      do 11 i=1,N
        write(*,'(1x,i2,2x,2f12.6)') i,x(i),f(i)
11    continue
      END

      SUBROUTINE funcv(n,x,f)
      INTEGER n
      REAL x(n),f(n)
```

```
f(1)=x(1)**2+x(2)**2-2.
f(2)=exp(x(1)-1.)+x(2)**3-2.
return
END
```

Chapter 10: Minimization and Maximization of Functions

Chapter 10 of *Numerical Recipes* deals with finding the maxima and minima of functions. The task has two parts, first the discovery of one or more bracketing intervals, and then the convergence to an extremum. mnbrak begins with two specified abscissas of a function and searches in the "downhill" direction for brackets of a minimum. golden can then take a bracketing triplet and perform a golden section search to a specified precision, for the minimum itself. When you are not concerned with worst-case examples, but only very efficient average-case performance, Brent's method (routine brent) is recommended. In the event that means are at hand for calculating the function's derivative as well as its value, consider dbrent.

Most multidimensional minimization strategies are based on the above one-dimensional algorithms. Our single example of an algorithm that is *not* so based is amoeba, which utilizes the downhill simplex method. Among the ones that do use one-dimensional methods are powell, frprmn, and dfpmin. The first two make calls to linmin, a subroutine that minimizes a function along a given direction in space. linmin in turn uses the one-dimensional algorithm brent, if derivatives are not known, or dbrent if they are. powell uses only function values and minimizes along an artfully chosen set of favorable directions. frprmn uses a Fletcher-Reeves-Polak-Ribiere minimization and requires the calculation of derivatives for the function. dfpmin uses a globally convergent variant of the Davidon-Fletcher-Powell variable metric method. This, too, requires calculation of derivatives, but the routine uses lnsrch for approximate one-dimensional minimizations.

The chapter ends with two topics of somewhat different nature. The first is linear programming, which deals with the maximization of a linear combination of variables, subject to linear constraints. This problem is dealt with by the simplex method in routine simplx. The second topic is the subject of large scale optimization, which is illustrated with the method of simulated annealing. One application of this is a combinatorial minimization problem, the "travelling salesman" problem in routine anneal. The second is the use of simulated annealing to find the global minimum in a continuous space, subroutine amebsa.

$\star$ $\quad$ $\star$ $\quad$ $\star$ $\quad$ $\star$

mnbrak searches a given function for a minimum. Given two values ax and bx of abscissa, it searches in the downward direction until it can find three new values `ax,bx,cx` that bracket a minimum. `fa,fb,fc` are the values of the function at these points. Sample program `xmnbrak` is a simple application of `mnbrak` applied to the Bessel function J_0. It tries a series of starting values `ax,bx` each encompassing an interval of length 1.0. `mnbrak` then finds several bracketing intervals of various minima of J_0.

```
      PROGRAM xmnbrak
C     driver for routine mnbrak
      INTEGER i
      REAL ax,bx,cx,fa,fb,fc
      EXTERNAL bessj0
      do 11 i=1,10
        ax=i*0.5
        bx=(i+1.0)*0.5
        call mnbrak(ax,bx,cx,fa,fb,fc,bessj0)
        write(*,'(1x,t13,a,t25,a,t37,a)') 'A','B','C'
        write(*,'(1x,a3,t5,3f12.6)') 'X',ax,bx,cx
        write(*,'(1x,a3,t5,3f12.6)') 'F',fa,fb,fc
11    continue
      END
```

Routine `golden` continues the minimization process by taking a bracketing triplet `ax,bx,cx` and performing a golden section search to isolate the contained minimum to a stated precision TOL. Sample program `xgolden` again uses J_0 as the test function. Using intervals (`ax,bx`) of length 1.0 it uses `mnbrak` to bracket all minima between $x = 0.0$ and $x = 100.0$. Some minima are bracketed more than once. On each pass, the bracketed solution is tracked down by `golden`. It is then compared to all previously located minima, and if different it is added to the collection by incrementing `nmin` (number of minima found) and adding the location `xmin` of the minima to the list in array `amin`. As a check of `golden`, the routine prints out the value of J_0 at the minimum, and also the value of J_1, which ought to be zero at extrema of J_0.

```
      PROGRAM xgolden
C     driver for routine golden
      REAL EQL,TOL
      PARAMETER(TOL=1.0E-6,EQL=1.E-3)
      INTEGER i,iflag,j,nmin
      REAL amin(20),ax,bx,cx,fa,fb,fc,g,xmin
      REAL bessj0,bessj1,golden
      EXTERNAL bessj0
      nmin=0
      write(*,'(/1x,a)') 'Minima of the function BESSJ0'
      write(*,'(/1x,t6,a,t19,a,t27,a,t40,a/)') 'Min. #','X',
     *     'BESSJ0(X)','BESSJ1(X)'
      do 12 i=1,100
        ax=i
        bx=i+1.0
        call mnbrak(ax,bx,cx,fa,fb,fc,bessj0)
        g=golden(ax,bx,cx,bessj0,TOL,xmin)
        if (nmin.eq.0) then
          amin(1)=xmin
          nmin=1
          write(*,'(1x,5x,i2,3x,3f12.6)') nmin,xmin,
     *          bessj0(xmin),bessj1(xmin)
        else
          iflag=0
```

```
          do 11 j=1,nmin
            if (abs(xmin-amin(j)).le.EQL*xmin)
     *            iflag=1
11        continue
          if (iflag.eq.0) then
            nmin=nmin+1
            amin(nmin)=xmin
            write(*,'(1x,5x,i2,3x,3f12.6)') nmin,
     *            xmin,bessj0(xmin),bessj1(xmin)
          endif
        endif
12      continue
        END
```

There are two other routines presented that also take the bracketing triplet ax,bx,cx from mnbrak and find the contained minimum. They are brent and dbrent. The sample programs for these two, xbrent and xdbrent, are virtually identical to that used on golden. Consequently, we can focus on the minor differences. For one thing, the driver for golden just passes the function name bessj0 to golden as an argument, and declares bessj0 as EXTERNAL to make this possible. In the program xbrent of brent, the function is called func, and func is declared EXTERNAL. Then a trivial FUNCTION subroutine is provided with func(x) set equal to bessj0(x). This illustrates two methods of accomplishing the same goal. The advantage of the second method is that if you are planning to deal with several different functions, you can avoid multiple compilations of the program; simply recompile the very short function subroutines and link each in turn to the same program. In program xdbrent we combined methods by passing bessj0 as an argument, and defining its derivative (−bessj1) in an external function routine deriv. Note that dbrent is only used when the derivative can be calculated conveniently.

```
        PROGRAM xbrent
C       driver for routine brent
        REAL EQL,TOL
        PARAMETER(TOL=1.0E-6,EQL=1.E-4)
        INTEGER i,iflag,j,nmin
        REAL amin(20),ax,b,bx,cx,fa,fb,fc,xmin
        REAL bessj0,bessj1,brent,func
        EXTERNAL func
        nmin=0
        write(*,'(/1x,a)') 'Minima of the function BESSJ0'
        write(*,'(/1x,t6,a,t19,a,t28,a,t40,a/)') 'Min. #','X',
     *       'BESSJ0(X)','BESSJ1(X)'
        do 12 i=1,100
          ax=i
          bx=i+1.0
          call mnbrak(ax,bx,cx,fa,fb,fc,func)
          b=brent(ax,bx,cx,func,TOL,xmin)
          if (nmin.eq.0) then
            amin(1)=xmin
            nmin=1
            write(*,'(1x,5x,i2,3x,3f12.6)') nmin,xmin,
     *            bessj0(xmin),bessj1(xmin)
          else
            iflag=0
            do 11 j=1,nmin
```

```
            if (abs(xmin-amin(j)).le.EQL*xmin)
     *              iflag=1
11          continue
            if (iflag.eq.0) then
              nmin=nmin+1
              amin(nmin)=xmin
              write(*,'(1x,5x,i2,3x,3f12.6)') nmin,
     *              xmin,bessj0(xmin),bessj1(xmin)
            endif
          endif
12      continue
        END

        REAL FUNCTION func(x)
        REAL bessj0,x
        func=bessj0(x)
        return
        END


        PROGRAM xdbrent
C       driver for routine dbrent
        REAL EQL,TOL
        PARAMETER(TOL=1.0E-6,EQL=1.E-4)
        INTEGER i,iflag,j,nmin
        REAL amin(20),ax,bx,cx,dbr,fa,fb,fc,xmin
        REAL bessj0,dbrent,deriv
        EXTERNAL bessj0,deriv
        nmin=0
        write(*,'(/1x,a)') 'Minima of the function BESSJ0'
        write(*,'(/1x,t6,a,t19,a,t27,a,t39,a,t53,a/)') 'Min. #','X',
     *      'BESSJ0(X)','BESSJ1(X)','DBRENT'
        do 12 i=1,100
          ax=i
          bx=i+1.0
          call mnbrak(ax,bx,cx,fa,fb,fc,bessj0)
          dbr=dbrent(ax,bx,cx,bessj0,deriv,TOL,xmin)
          if (nmin.eq.0) then
            amin(1)=xmin
            nmin=1
            write(*,'(1x,5x,i2,3x,4f12.6)') nmin,xmin,
     *            bessj0(xmin),deriv(xmin),dbr
          else
            iflag=0
            do 11 j=1,nmin
              if (abs(xmin-amin(j)).le.EQL*xmin) iflag=1
11          continue
            if (iflag.eq.0) then
              nmin=nmin+1
              amin(nmin)=xmin
              write(*,'(1x,5x,i2,3x,4f12.6)') nmin,xmin,
     *              bessj0(xmin),deriv(xmin),dbr
            endif
          endif
12      continue
        END

        REAL FUNCTION deriv(x)
```

```
REAL bessj1,x
deriv=-bessj1(x)
END
```

Numerical Recipes presents several methods for minimization in multiple dimensions. Among these, the downhill simplex method carried out by amoeba is the only one that does not treat the problem as a series of one-dimensional minimizations. As input, amoeba requires the coordinates of $N + 1$ vertices of a starting simplex in N-dimensional space, and the values y of the function at each of these vertices. Sample program xamoeba tries the method out on the exotic function

$$\text{famoeb} = 0.6 - J_0[(x - 0.5)^2 + (y - 0.6)^2 + (z - 0.7)^2]$$

which has a minimum at $(x, y, z) = (0.5, 0.6, 0.7)$. As vertices of the starting simplex, specified by p in a DATA statement, we used $(0, 0, 0)$, $(1, 0, 0)$, $(0, 1, 0)$, and $(0, 0, 1)$. A vector x(i) is set successively to each vertex to allow the evaluation of function values y. This data is submitted to amoeba along with FTOL=1.0E-6 to specify the tolerance on the function value. The vertices and corresponding function values of the final simplex are printed out, and you can easily check whether the specified tolerance is met.

```
      PROGRAM xamoeba
C     driver for routine amoeba
      INTEGER NP,MP
      REAL FTOL
      PARAMETER(NP=3,MP=4,FTOL=1.0E-6)
      INTEGER i,iter,j,ndim
      REAL famoeb,p(MP,NP),x(NP),y(MP)
      EXTERNAL famoeb
      DATA p/0.0,1.0,0.0,0.0,0.0,0.0,1.0,0.0,0.0,0.0,0.0,1.0/
      ndim=NP
      do 12 i=1,MP
        do 11 j=1,NP
          x(j)=p(i,j)
11      continue
        y(i)=famoeb(x)
12    continue
      call amoeba(p,y,MP,NP,ndim,FTOL,famoeb,iter)
      write(*,'(/1x,a,i3)') 'Number of iterations: ',iter
      write(*,'(/1x,a)') 'Vertices of final 3-D simplex and'
      write(*,'(1x,a)') 'function values at the vertices:'
      write(*,'(/3x,a,t11,a,t23,a,t35,a,t45,a/)') 'I',
     *    'X(I)','Y(I)','Z(I)','FUNCTION'
      do 13 i=1,MP
        write(*,'(1x,i3,4f12.6)') i,(p(i,j),j=1,NP),y(i)
13    continue
      write(*,'(/1x,a)') 'True minimum is at (0.5,0.6,0.7)'
      END

      REAL FUNCTION famoeb(x)
      REAL bessj0,x(3)
      famoeb=0.6-bessj0((x(1)-0.5)**2+(x(2)-0.6)**2+(x(3)-0.7)**2)
      END
```

powell carries out one-dimensional minimizations along favorable directions in N-dimensional space. The function minimized must be called func, and in sample program `xpowell` a function subroutine is defined for

$$\text{func}(x,y,z) = \tfrac{1}{2} - J_0[(x-1)^2 + (y-2)^2 + (z-3)^2].$$

The program provides `powell` with a starting point P of $(3/2, 3/2, 5/2)$ and a set of initial directions, here chosen to be the unit directions $(1,0,0)$, $(0,1,0)$, and $(0,0,1)$. `powell` performs its one-dimensional minimizations with `linmin`, which is discussed next.

```
      PROGRAM xpowell
C     driver for routine powell
      INTEGER NDIM
      REAL FTOL
      PARAMETER(NDIM=3,FTOL=1.0E-6)
      INTEGER i,iter,np
      REAL fret,p(NDIM),xi(NDIM,NDIM)
      np=NDIM
      DATA xi/1.0,0.0,0.0,0.0,1.0,0.0,0.0,0.0,1.0/
      DATA p/1.5,1.5,2.5/
      call powell(p,xi,NDIM,np,FTOL,iter,fret)
      write(*,'(/1x,a,i3)') 'Iterations:',iter
      write(*,'(/1x,a/1x,3f12.6)') 'Minimum found at: ',(p(i),i=1,NDIM)
      write(*,'(/1x,a,f12.6)') 'Minimum function value =',fret
      write(*,'(/1x,a)') 'True minimum of function is at:'
      write(*,'(1x,3f12.6/)') 1.0,2.0,3.0
      END

      REAL FUNCTION func(x)
      REAL bessj0,x(3)
      func=0.5-bessj0((x(1)-1.0)**2+(x(2)-2.0)**2+(x(3)-3.0)**2)
      END
```

`linmin`, as we have said, finds the minimum of a function `func` along a direction in N-dimensional space. To use it we specify a point P and a direction vector `xi`, both in N-space. `linmin` then does the bookkeeping required to treat the function as a function of position along this line, and minimizes the function with a conventional one-dimensional minimization routine. Sample program `xlinmin` feeds `linmin` the function

$$\text{func}(x,y,z) = (x-1)^2 + (y-1)^2 + (z-1)^2$$

which has a minimum at $(x,y,z) = (1,1,1)$. It also chooses point P to be the origin $(0,0,0)$, and tries a series of directions

$$\left(\sqrt{2}\cos\left(\frac{\pi}{2}\frac{i}{10.0}\right), \quad \sqrt{2}\sin\left(\frac{\pi}{2}\frac{i}{10.0}\right), \quad 1.0 \right) \qquad i = 1,\ldots,10$$

For each pass, the location of the minimum, and the value of the function at the minimum, are printed. Among the directions searched is the direction $(1,1,1)$. Along this direction, of course, the minimum function value should be zero and should occur at $(1,1,1)$.

```
      PROGRAM xlinmin
C     driver for routine linmin
      INTEGER NDIM
      REAL PIO2
      PARAMETER(NDIM=3,PIO2=1.5707963)
```

```
      INTEGER i,j
      REAL fret,sr2,x,p(NDIM),xi(NDIM)
      write(*,'(/1x,a)') 'Minimum of a 3-D quadratic centered'
      write(*,'(1x,a)') 'at (1.0,1.0,1.0). Minimum is found'
      write(*,'(1x,a)') 'along a series of radials.'
      write(*,'(/1x,t10,a,t22,a,t34,a,t42,a/)') 'x','y','z','minimum'
      do 11 i=0,10
        x=PIO2*i/10.0
        sr2=sqrt(2.0)
        xi(1)=sr2*cos(x)
        xi(2)=sr2*sin(x)
        xi(3)=1.0
        p(1)=0.0
        p(2)=0.0
        p(3)=0.0
        call linmin(p,xi,NDIM,fret)
        write(*,'(1x,4f12.6)') (p(j),j=1,3),fret
11    continue
      END


      REAL FUNCTION func(x)
      INTEGER i
      REAL x(3)
      func=0.0
      do 11 i=1,3
        func=func+(x(i)-1.0)**2
11    continue
      END
```

f1dim accompanies `linmin` and is the routine that makes the N-dimensional function `func` effectively a one-dimensional function along a given line in N-space. There is little to check here, and our perfunctory demonstration of its use, in sample program `xf1dim`, simply plots `f1dim` as a one dimensional function, given the function

$$\text{func}(x,y,z) = (x-1)^2 + (y-1)^2 + (z-1)^2.$$

You get to choose the direction; then `scrsho` plots the function along this direction. Try the direction $(1,1,1)$ along which you should find a minimum value of `func`=0 at position $(1,1,1)$.

```
      PROGRAM xf1dim
C     driver for routine f1dim
      INTEGER NDIM,NMAX
      PARAMETER(NDIM=3,NMAX=50)
      INTEGER i,j,ncom
      REAL pcom,xicom
      COMMON /f1com/ pcom(NMAX),xicom(NMAX),ncom
      REAL p(NDIM),xi(NDIM)
      EXTERNAL f1dim
      DATA p/0.0,0.0,0.0/
      ncom=NDIM
      write(*,'(/1x,a)') 'Enter vector direction along which to'
      write(*,'(1x,a)') 'plot the function. Minimum is in the'
      write(*,'(1x,a)') 'direction 1.0,1.0,1.0 - Enter X,Y,Z:'
      read(*,*) (xi(i),i=1,3)
      do 11 j=1,NDIM
        pcom(j)=p(j)
```

```
            xicom(j)=xi(j)
11      continue
        call scrsho(f1dim)
        END

        REAL FUNCTION func(x)
        INTEGER i
        REAL x(3)
        func=0.0
        do 11 i=1,3
            func=func+(x(i)-1.0)**2
11      continue
        END
```

frprmn is another multidimensional minimizer that relies on the one-dimensional minimizations of linmin. It works, however, via the Fletcher-Reeves-Polak-Ribiere method and requires that routines be supplied for calculating both the function and its gradient. Sample program xfrprmn, for example, uses

$$\text{func}(x,y,z) = 1.0 - J_0(x - \tfrac{1}{2})J_0(y - \tfrac{1}{2})J_0(z - \tfrac{1}{2})$$

and

$$\frac{\partial\,\text{func}}{\partial x} = J_1(x - \tfrac{1}{2})J_0(y - \tfrac{1}{2})J_0(z - \tfrac{1}{2})$$

etc. A number of trial starting vectors are used, and each time, frprmn manages to find the minimum at $(1/2, 1/2, 1/2)$.

```
        PROGRAM xfrprmn
C       driver for routine frprmn
        INTEGER NDIM
        REAL FTOL,PIO2
        PARAMETER(NDIM=3,FTOL=1.0E-6,PIO2=1.5707963)
        INTEGER iter,k
        REAL angl,fret,p(NDIM)
        write(*,'(/1x,a)') 'Program finds the minimum of a function'
        write(*,'(1x,a)') 'with different trial starting vectors.'
        write(*,'(1x,a)') 'True minimum is (0.5,0.5,0.5)'
        do 11 k=0,4
            angl=PIO2*k/4.0
            p(1)=2.0*cos(angl)
            p(2)=2.0*sin(angl)
            p(3)=0.0
            write(*,'(/1x,a,3(f6.4,a))') 'Starting vector: (',
     *          p(1),',',p(2),',',p(3),')'
            call frprmn(p,NDIM,FTOL,iter,fret)
            write(*,'(1x,a,i3)') 'Iterations:',iter
            write(*,'(1x,a,3(f6.4,a))') 'Solution vector: (',
     *          p(1),',',p(2),',',p(3),')'
            write(*,'(1x,a,e14.6)') 'Func. value at solution',fret
11      continue
        END

        REAL FUNCTION func(x)
        REAL bessj0,x(3)
        func=1.0-bessj0(x(1)-0.5)*bessj0(x(2)-0.5)*bessj0(x(3)-0.5)
        END
```

```
     SUBROUTINE dfunc(x,df)
     INTEGER NMAX
     PARAMETER (NMAX=50)
     REAL bessj0,bessj1,x(3),df(NMAX)
     df(1)=bessj1(x(1)-0.5)*bessj0(x(2)-0.5)*bessj0(x(3)-0.5)
     df(2)=bessj0(x(1)-0.5)*bessj1(x(2)-0.5)*bessj0(x(3)-0.5)
     df(3)=bessj0(x(1)-0.5)*bessj0(x(2)-0.5)*bessj1(x(3)-0.5)
     return
     END
```

Completeness requires that we provide a sample program for df1dim, which is presented in *Numerical Recipes* as a routine for converting the N-dimensional gradient subroutine to one that provides the first derivative of the function along a specified line in N-dimensional space. It is exactly analogous to f1dim and the program xdf1dim is the same.

```
     PROGRAM xdf1dim
C    driver for routine df1dim
     INTEGER NDIM,NMAX
     PARAMETER(NDIM=3,NMAX=50)
     INTEGER i,j,ncom
     REAL pcom,xicom
     COMMON /f1com/ pcom(NMAX),xicom(NMAX),ncom
     REAL df1dim,p(NDIM),xi(NDIM)
     EXTERNAL df1dim
     DATA p/0.0,0.0,0.0/
     ncom=NDIM
     write(*,'(/1x,a)') 'Enter vector direction along which to'
     write(*,'(1x,a)') 'plot the function. Minimum is in the'
     write(*,'(1x,a)') 'direction 1.0,1.0,1.0 - Enter X,Y,Z:'
     read(*,*) (xi(i),i=1,3)
     do 11 j=1,NDIM
       pcom(j)=p(j)
       xicom(j)=xi(j)
11   continue
     call scrsho(df1dim)
     END

     SUBROUTINE dfunc(x,df)
     INTEGER i
     REAL x(3),df(3)
     do 11 i=1,3
       df(i)=(x(i)-1.0)**2
11   continue
     return
     END
```

dfpmin implements the Broyden-Fletcher-Goldfarb-Shanno variant of the David-on-Fletcher-Powell minimization by variable metric methods. It requires somewhat more intermediate storage than the preceding routine, but it works well with a globally convergent strategy that requires only approximate line minimizations, and can therefore be more efficient. Sample program xdfpmin works just as did the program for frprmn, including the fact that it requires a subroutine for calculation of the derivative. The approximate line minimizations are carried out by lnsrch. In this case we try it on the somewhat harder problem (Stoer and Bulirsch, *Introduction to Numerical Analysis*, p. 308):

$$f = 10[(x_2^2(3 - x_1) - x_1^2(3 + x_1))]^2 + \frac{(2 + x_1)^2}{1 + (2 + x_1)^2}$$

which has a minimum at $(x_1, x_2) = (-2, \pm 0.89442719)$. We take the initial guess to be $(0.1, 4.2)$. (The original problem in Stoer and Bulirsch has 100 in the first term instead of 10. We have changed it because some machines may have trouble in single precision. Even this easier problem requires double precision if you try to solve it with powell.)

```
      PROGRAM xdfpmin
C     driver for routine dfpmin
      INTEGER NDIM
      REAL GTOL
      PARAMETER(NDIM=2,GTOL=1.0E-4)
      COMMON /stats/ nfunc,ndfunc
      INTEGER iter,ndfunc,nfunc
      REAL fret,p(NDIM)
      EXTERNAL func,dfunc
      write(*,'(/1x,a)') 'True minimum is at (-2.0,+-0.89442719)'
      nfunc=0
      ndfunc=0
      p(1)=.1
      p(2)=4.2
      write(*,'(/1x,a,2(f7.4,a))') 'Starting vector: (',
     *     p(1),',',p(2),')'
      call dfpmin(p,NDIM,GTOL,iter,fret,func,dfunc)
      write(*,'(1x,a,i3)') 'Iterations:',iter
      write(*,'(1x,a,i3)') 'Func. evals:',nfunc
      write(*,'(1x,a,i3)') 'Deriv. evals:',ndfunc
      write(*,'(1x,a,2(f9.6,a))') 'Solution vector: (',
     *     p(1),',',p(2),')'
      write(*,'(1x,a,e14.6)') 'Func. value at solution',fret
      END

      REAL FUNCTION func(x)
      INTEGER ndfunc,nfunc
      COMMON /stats/ nfunc,ndfunc
      REAL x(*)
      nfunc=nfunc+1
      func=10.*(x(2)**2*(3.-x(1))-x(1)**2*(3.+x(1)))**2+(2.+x(1))**2/
     *     (1.+(2.+x(1))**2)
      END

      SUBROUTINE dfunc(x,df)
      INTEGER NMAX
      PARAMETER (NMAX=50)
      INTEGER ndfunc,nfunc
      COMMON /stats/ nfunc,ndfunc
      REAL x(*),df(NMAX)
      ndfunc=ndfunc+1
      df(1)=20.*(x(2)**2*(3.-x(1))-x(1)**2*(3.+x(1)))*(-x(2)**2-6.*
     *     x(1)-3.*x(1)**2)+2.*(2.+x(1))/(1.+(2.+x(1))**2)-
     *     2.*(2.+x(1))**3/(1.+(2.+x(1))**2)**2
      df(2)=40.*(x(2)**2*(3.-x(1))-x(1)**2*(3.+x(1)))*x(2)*(3.-x(1))
```

```
      return
      END
```

simplx is a subroutine for dealing with problems in linear programming. In these problems the goal is to maximize a linear combination of N variables, subject to the constraint that none be negative, and that as a group they satisfy a number of other constraints. In order to clarify the subject, *Numerical Recipes* presents a sample problem in equations (10.8.6) and (10.8.7), translating the problem into tableau format in (10.8.18), and presenting a solution in equation (10.8.19). Sample program xsimplx carries out the calculations that lead to this solution.

```
      PROGRAM xsimplx
C     driver for routine simplx
      INTEGER N,M,NP,MP,M1,M2,M3,NM1M2
      PARAMETER(N=4,M=4,NP=5,MP=6,M1=2,M2=1,M3=1,NM1M2=N+M1+M2)
      INTEGER i,icase,j,jj,jmax,izrov(N),iposv(M)
      REAL a(MP,NP),anum(NP)
      CHARACTER txt(NM1M2)*2,alpha(NP)*2
      LOGICAL rite
      DATA txt/'x1','x2','x3','x4','y1','y2','y3'/
      DATA a/0.0,740.0,0.0,0.5,9.0,0.0,1.0,-1.0,0.0,0.0,-1.0,0.0,
     *      1.0,0.0,-2.0,-1.0,-1.0,0.0,3.0,-2.0,0.0,1.0,-1.0,0.0,
     *      -0.5,0.0,7.0,-2.0,-1.0,0.0/
      call simplx(a,M,N,MP,NP,M1,M2,M3,icase,izrov,iposv)
      if (icase.eq.1) then
        write(*,*) 'Unbounded objective function'
      else if (icase.eq.-1) then
        write(*,*) 'No solutions satisfy constraints given'
      else
        jj=1
        do 11 i=1,N
          if (izrov(i).le.NM1M2) then
            alpha(jj)=txt(izrov(i))
            jj=jj+1
          endif
11      continue
        jmax=jj-1
        write(*,'(/3x,5a10)') ' ',(alpha(jj),jj=1,jmax)
        do 13 i=1,M+1
          if (i.eq.1) then
            alpha(1)=' '
            rite=.true.
          else if (iposv(i-1).le.NM1M2) then
            alpha(1)=txt(iposv(i-1))
            rite=.true.
          else
            rite=.false.
          endif
          if (rite) then
            anum(1)=a(i,1)
            jj=2
            do 12 j=2,N+1
              if (izrov(j-1).le.NM1M2) then
                anum(jj)=a(i,j)
                jj=jj+1
              endif
```

```
12        continue
          jmax=jj-1
          write(*,'(1x,a3,(5f10.2))') alpha(1),(anum(jj),jj=1,jmax)
        endif
13      continue
      endif
      END
```

anneal is a subroutine for solving the travelling salesman problem — a problem
that is included as a demonstration of the use of simulated annealing. Sample program
xanneal has the function of setting up the initial route for the salesman and printing
final results. For each of NCITY=10 cities, it chooses random coordinates x(i),y(i)
using routine ran3, and puts an entry for each city in the array iptr(i). The array
indicates the order in which the cities will be visited. On the originally specified path,
the cities are in the order i=1,..,10 so the sample program initially takes iptr(i)=i.
(It is assumed that the salesman will return to the first city after visiting the last.)
A call is then made to anneal, which attempts to find the shortest alternative route,
which is recorded in the array. After finding a path that resists further improvement,
the driver lists the modified itinerary.

```
      PROGRAM xanneal
C     driver for routine anneal
      INTEGER NCITY
      PARAMETER (NCITY=10)
      INTEGER i,idum,ii,iorder(NCITY)
      REAL ran3,x(NCITY),y(NCITY)
C     create points of sale
      idum=-111
      do 11 i=1,NCITY
        x(i)=ran3(idum)
        y(i)=ran3(idum)
        iorder(i)=i
11      continue
      call anneal(x,y,iorder,NCITY)
      write(*,*) '*** System Frozen ***'
      write(*,*) 'Final path:'
      write(*,'(1x,t3,a,t13,a,t23,a)') 'city','x','y'
      do 12 i=1,NCITY
        ii=iorder(i)
        write(*,'(1x,i4,2f10.4)') ii,x(ii),y(ii)
12      continue
      END
```

If you have any doubts about the effectiveness of simulated annealing in dealing
with otherwise intractable minimization problems with continuous control variables,
the next example, using routine amebsa, ought to convince you of the merits of
the technique. The 4-dimensional test function to be minimized, tfunk, is quite
diabolical: It is bounded below by a parabolic function whose principal axes are in
the ratio 1 : 3 : 10 : 30, and bounded above by 3 (or more generally 1+AUG) times
the lower bound. At most points, the function in fact takes on its upper bound; but
within a radius 0.3 (that is, RAD) of any *integer* lattice point, it dives down to its lower
bound in a narrow parabola. The narrow parabolas, in other words, form an infinite
lattice of local minimum "traps," trying to capture the test point. The "surface" of
this function would look like a piece of 4-dimensional Swiss cheese, scooped out into a

paraboloid. Any minimization algorithm that goes strictly downhill will get captured by the holes, virtually always. Let's see how well simulated annealing does at finding the true minimum, located at the origin.

The program prompts for a starting temperature, T, and for the number of amoeba iterations, ITER, that it should perform before reducing the temperature by 20%. The latter number, therefore, controls the annealing schedule. The test point is alway started at a position $\mathbf{p} = (10, 10, 10, 10)$, down at the bottom of a steep local minimum. Note how the program sets up the initial simplex and its function values, as required by the routine amebsa.

Try running the program first with T=1000, ITER=20. For this low a value of T, the routine never gets out of its original local minimum, and it ends up with a function value of 101001. Next try the input values T=1×10^6, ITER=20. For most random seeds, the program now finds its way to the origin, with a final function value of 1, the true global minimum.

To see the effect of too rapid quenching, try T=1×10^6, ITER=2. For many random seeds, the program now gets part of the way towards the origin but then freezes into a local minimum. For other seeds, the program does find the true global minimum. You might enjoy experimenting with other input values to map out the regions of relatively secure and insecure global convergence.

```
      PROGRAM xamebsa
C     driver for routine amebsa
      INTEGER NP,MP,MPNP
      REAL FTOL
      PARAMETER(NP=4,MP=5,MPNP=20,FTOL=1.0E-6)
      INTEGER i,idum,iiter,iter,j,jiter,ndim,nit
      REAL temptr,tt,yb,ybb
      COMMON /ambsa/ tt,idum
      REAL p(MP,NP),x(NP),y(MP),xoff(NP),pb(NP),tfunk
      EXTERNAL tfunk
      DATA xoff/4*10./
      DATA p/MPNP*0.0/
      idum=-64
1     continue
      ndim=NP
      do 11 j=2,MP
        p(j,j-1)=1.
11    continue
      do 13 i=1,MP
        do 12 j=1,NP
          p(i,j)=p(i,j)+xoff(j)
          x(j)=p(i,j)
12      continue
        y(i)=tfunk(x)
13    continue
      yb=1.e30
      write(*,*) 'Input T, IITER:'
      read(*,*,END=999) temptr,iiter
      ybb=1.e30
      nit=0
      do 14 jiter=1,100
        iter=iiter
        temptr=temptr*0.8
```

```
        call amebsa(p,y,MP,NP,ndim,pb,yb,FTOL,tfunk,iter,temptr)
        nit=nit+iiter-iter
        if (yb.lt.ybb) then
          ybb=yb
          write(*,'(1x,i6,e10.3,4f11.5,e15.7)')')
     *        nit,temptr,(pb(j),j=1,NP),yb
        endif
        if (iter.gt.0) goto 80
14      continue
80      write(*,'(/1x,a)') 'Vertices of final 3-D simplex and'
        write(*,'(1x,a)') 'function values at the vertices:'
        write(*,'(/3x,a,t11,a,t23,a,t35,a,t45,a/)') 'I',
     *      'X(I)','Y(I)','Z(I)','FUNCTION'
        do 15 i=1,MP
          write(*,'(1x,i3,4f12.6,e15.7)') i,(p(i,j),j=1,NP),y(i)
15      continue
        write(*,'(1x,i3,4f12.6,e15.7)') 99,(pb(j),j=1,NP),yb
        goto 1
999     write(*,*) 'NORMAL COMPLETION'
        STOP
        END

        REAL FUNCTION tfunk(p)
        INTEGER N
        REAL RAD,AUG
        PARAMETER (N=4,RAD=0.3,AUG=2.0)
        INTEGER j
        REAL q,r,sumd,sumr,p(N),wid(N)
        DATA wid /1.,3.,10.,30./
        sumd=0.
        sumr=0.
        do 11 j=1,N
          q=p(j)*wid(j)
          r=nint(q)
          sumr=sumr+q**2
          sumd=sumd+(q-r)**2
11      continue
        if (sumd.gt.RAD**2) then
          tfunk=sumr*(1.+AUG)+1.
        else
          tfunk=sumr*(1.+AUG*sumd/RAD**2)+1.
        endif
        return
        END
```

Chapter 11: Eigensystems

In *Chapter 11 of Numerical Recipes, we deal with the problem of finding eigenvectors and eigenvalues of matrices, first dealing with symmetric matrices, and then with more general cases. For real symmetric matrices of small-to-moderate size, the routine* jacobi *is recommended as a simple and foolproof scheme of finding eigenvalues and eigenvectors. Routine* eigsrt *may be used to reorder the output of* jacobi *into descending order of eigenvalue. A more efficient (but operationally more complicated) procedure is to reduce the symmetric matrix to tridiagonal form before doing the eigenvalue analysis.* tred2 *uses the Householder scheme to perform this reduction and is used in conjunction with* tqli. tqli *determines the eigenvalues and eigenvectors of a real, symmetric, tridiagonal matrix.*

For nonsymmetric matrices, we offer only routines for finding eigenvalues, and not eigenvectors. To ameliorate problems with roundoff error, balanc *makes the corresponding rows and columns of the matrix have comparable norms while leaving eigenvalues unchanged. Then the matrix is reduced to Hessenberg form by Gaussian elimination using* elmhes. *Finally* hqr *applies the QR algorithm to find the eigenvalues of the Hessenberg matrix.*

$\star \quad \star \quad \star \quad \star$

jacobi is a reliable scheme for finding both the eigenvalues and eigenvectors of a symmetric matrix. It is not the most efficient scheme available, but it is simple and trustworthy, and it is recommended for problems of small-to-moderate order. Sample program xjacobi defines three matrices a,b,c for use by jacobi. They are of order 3, 5, and 10 respectively. To check the index handling in jacobi, they are, each in turn, loaded into a 10×10 matrix e, and then sent to jacobi with NP=10 (the physical size of the matrix) and n=3, 5 or 10 (the logical array dimension). For each matrix, the eigenvalues and eigenvectors are reported. Then, an eigenvector test takes place in which the original matrix is applied to the purported eigenvector, and the ratio of the result to the vector itself is found. The ratio should, of course, be the eigenvalue.

```
      PROGRAM xjacobi
C     driver for routine jacobi
      INTEGER NP,NMAT
      PARAMETER(NP=10,NMAT=3)
      INTEGER i,ii,j,jj,k,kk,l,ll,nrot,num(3)
      REAL ratio,d(NP),v(NP,NP),r(NP)
      REAL a(3,3),b(5,5),c(10,10),e(NP,NP)
      DATA num/3,5,10/
      DATA a/1.0,2.0,3.0,2.0,2.0,3.0,3.0,3.0,3.0/
      DATA b/-2.0,-1.0,0.0,1.0,2.0,-1.0,-1.0,0.0,1.0,2.0,
```

```
      *     0.0,0.0,0.0,1.0,2.0,1.0,1.0,1.0,1.0,2.0,
      *     2.0,2.0,2.0,2.0,2.0/
      DATA c/5.0,4.3,3.0,2.0,1.0,0.0,-1.0,-2.0,-3.0,-4.0,
      *     4.3,5.1,4.0,3.0,2.0,1.0,0.0,-1.0,-2.0,-3.0,
      *     3.0,4.0,5.0,4.0,3.0,2.0,1.0,0.0,-1.0,-2.0,
      *     2.0,3.0,4.0,5.0,4.0,3.0,2.0,1.0,0.0,-1.0,
      *     1.0,2.0,3.0,4.0,5.0,4.0,3.0,2.0,1.0,0.0,
      *     0.0,1.0,2.0,3.0,4.0,5.0,4.0,3.0,2.0,1.0,
      *     -1.0,0.0,1.0,2.0,3.0,4.0,5.0,4.0,3.0,2.0,
      *     -2.0,-1.0,0.0,1.0,2.0,3.0,4.0,5.0,4.0,3.0,
      *     -3.0,-2.0,-1.0,0.0,1.0,2.0,3.0,4.0,5.0,4.0,
      *     -4.0,-3.0,-2.0,-1.0,0.0,1.0,2.0,3.0,4.0,5.0/
      do 24 i=1,NMAT
        if (i.eq.1) then
          do 12 ii=1,3
            do 11 jj=1,3
              e(ii,jj)=a(ii,jj)
11          continue
12        continue
          call jacobi(e,3,NP,d,v,nrot)
        else if (i.eq.2) then
          do 14 ii=1,5
            do 13 jj=1,5
              e(ii,jj)=b(ii,jj)
13          continue
14        continue
          call jacobi(e,5,NP,d,v,nrot)
        else if (i.eq.3) then
          do 16 ii=1,10
            do 15 jj=1,10
              e(ii,jj)=c(ii,jj)
15          continue
16        continue
          call jacobi(e,10,NP,d,v,nrot)
        endif
        write(*,'(/1x,a,i2)') 'Matrix Number ',i
        write(*,'(1x,a,i3)') 'Number of JACOBI rotations: ',nrot
        write(*,'(/1x,a)') 'Eigenvalues:'
        do 17 j=1,num(i)
          write(*,'(1x,5f12.6)') d(j)
17      continue
        write(*,'(/1x,a)') 'Eigenvectors:'
        do 18 j=1,num(i)
          write(*,'(1x,t5,a,i3)') 'Number',j
          write(*,'(1x,5f12.6)') (v(k,j),k=1,num(i))
18      continue
C     eigenvector test
        write(*,'(/1x,a)') 'Eigenvector Test'
        do 23 j=1,num(i)
          do 21 l=1,num(i)
            r(l)=0.0
            do 19 k=1,num(i)
              if (k.gt.l) then
                kk=l
                ll=k
              else
                kk=k
```

```
              ll=1
            endif
            if (i.eq.1) then
              r(1)=r(1)+a(ll,kk)*v(k,j)
            else if (i.eq.2) then
              r(1)=r(1)+b(ll,kk)*v(k,j)
            else if (i.eq.3) then
              r(1)=r(1)+c(ll,kk)*v(k,j)
            endif
19        continue
21      continue
        write(*,'(/1x,a,i3)') 'Vector Number',j
        write(*,'(/1x,t7,a,t18,a,t31,a)')
     *        'Vector','Mtrx*Vec.','Ratio'
        do 22 l=1,num(i)
          ratio=r(1)/v(1,j)
          write(*,'(1x,3f12.6)') v(1,j),r(1),ratio
22      continue
23    continue
      write(*,*) 'press RETURN to continue...'
      read(*,*)
24  continue
    END
```

eigsrt reorders the output of jacobi so that the eigenvectors are in the order of decreasing eigenvalue. Sample program xeigsrt uses matrix c from the previous program to illustrate. This 10×10 matrix is passed to jacobi and the ten eigenvectors are found. They are printed, along with their eigenvalues, in the order that jacobi returns them. Then the matrices d and v from jacobi, which contain the eigenvalues and eigenvectors, are passed to eigsrt, and ought to return in descending order of eigenvalue. The result is printed for inspection.

```
      PROGRAM xeigsrt
C     driver for routine eigsrt
      INTEGER NP
      PARAMETER(NP=10)
      INTEGER i,j,nrot
      REAL d(NP),v(NP,NP),c(NP,NP)
      DATA c /5.0,4.3,3.0,2.0,1.0,0.0,-1.0,-2.0,-3.0,-4.0,
     *     4.3,5.1,4.0,3.0,2.0,1.0,0.0,-1.0,-2.0,-3.0,
     *     3.0,4.0,5.0,4.0,3.0,2.0,1.0,0.0,-1.0,-2.0,
     *     2.0,3.0,4.0,5.0,4.0,3.0,2.0,1.0,0.0,-1.0,
     *     1.0,2.0,3.0,4.0,5.0,4.0,3.0,2.0,1.0,0.0,
     *     0.0,1.0,2.0,3.0,4.0,5.0,4.0,3.0,2.0,1.0,
     *     -1.0,0.0,1.0,2.0,3.0,4.0,5.0,4.0,3.0,2.0,
     *     -2.0,-1.0,0.0,1.0,2.0,3.0,4.0,5.0,4.0,3.0,
     *     -3.0,-2.0,-1.0,0.0,1.0,2.0,3.0,4.0,5.0,4.0,
     *     -4.0,-3.0,-2.0,-1.0,0.0,1.0,2.0,3.0,4.0,5.0/
      call jacobi(c,NP,NP,d,v,nrot)
      write(*,*) 'Unsorted Eigenvectors:'
      do 11 i=1,NP
        write(*,'(/1x,a,i3,a,f12.6)') 'Eigenvalue',i,' =',d(i)
        write(*,*) 'Eigenvector:'
        write(*,'(10x,5f12.6)') (v(j,i),j=1,NP)
11    continue
      write(*,'(//,1x,a,//)') '****** sorting ******'
```

```
      call eigsrt(d,v,NP,NP)
      write(*,*) 'Sorted Eigenvectors:'
      do 12 i=1,NP
        write(*,'(/1x,a,i3,a,f12.6)') 'Eigenvalue',i,' =',d(i)
        write(*,*) 'Eigenvector:'
        write(*,'(10x,5f12.6)') (v(j,i),j=1,NP)
12    continue
      END
```

tred2 reduces a real symmetric matrix to tridiagonal form. Sample program xtred2 again uses matrix c from the earlier programs, and copies it into matrix a. Matrix a is sent to tred2, while c is saved for a check of the transformation matrix that tred2 returns in a. The program prints the diagonal and off-diagonal elements of the reduced matrix. It then forms the matrix f defined by $F = A^T C A$ to prove that f is tridiagonal and that the listed diagonal and off-diagonal elements are correct.

```
      PROGRAM xtred2
C     driver for routine tred2
      INTEGER NP
      PARAMETER(NP=10)
      INTEGER i,j,k,l,m
      REAL a(NP,NP),c(NP,NP),d(NP),e(NP),f(NP,NP)
      DATA c/5.0,4.3,3.0,2.0,1.0,0.0,-1.0,-2.0,-3.0,-4.0,
     *     4.3,5.1,4.0,3.0,2.0,1.0,0.0,-1.0,-2.0,-3.0,
     *     3.0,4.0,5.0,4.0,3.0,2.0,1.0,0.0,-1.0,-2.0,
     *     2.0,3.0,4.0,5.0,4.0,3.0,2.0,1.0,0.0,-1.0,
     *     1.0,2.0,3.0,4.0,5.0,4.0,3.0,2.0,1.0,0.0,
     *     0.0,1.0,2.0,3.0,4.0,5.0,4.0,3.0,2.0,1.0,
     *     -1.0,0.0,1.0,2.0,3.0,4.0,5.0,4.0,3.0,2.0,
     *     -2.0,-1.0,0.0,1.0,2.0,3.0,4.0,5.0,4.0,3.0,
     *     -3.0,-2.0,-1.0,0.0,1.0,2.0,3.0,4.0,5.0,4.0,
     *     -4.0,-3.0,-2.0,-1.0,0.0,1.0,2.0,3.0,4.0,5.0/
      do 12 i=1,NP
        do 11 j=1,NP
          a(i,j)=c(i,j)
11      continue
12    continue
      call tred2(a,NP,NP,d,e)
      write(*,'(/1x,a)') 'Diagonal elements'
      write(*,'(1x,5f12.6)') (d(i),i=1,NP)
      write(*,'(/1x,a)') 'Off-diagonal elements'
      write(*,'(1x,5f12.6)') (e(i),i=2,NP)
C     check transformation matrix
      do 16 j=1,NP
        do 15 k=1,NP
          f(j,k)=0.0
          do 14 l=1,NP
            do 13 m=1,NP
              f(j,k)=f(j,k)
     *                +a(l,j)*c(l,m)*a(m,k)
13          continue
14        continue
15      continue
16    continue
C     how does it look?
      write(*,'(/1x,a)') 'Tridiagonal matrix'
```

```
      do 17 i=1,NP
        write(*,'(1x,10f7.2)') (f(i,j),j=1,NP)
17    continue
      END
```

tqli finds the eigenvectors and eigenvalues for a real, symmetric, tridiagonal matrix. Sample program xtqli operates with matrix c again, and uses tred2 to reduce it to tridiagonal form as before. More specifically, c is copied into matrix a, which is sent to tred2. From tred2 come two vectors d,e which are the diagonal and subdiagonal elements of the tridiagonal matrix. d and e are made arguments of tqli, as is a, the returned transformation matrix from tred2. On output from tqli, d is replaced with eigenvalues, and a with corresponding eigenvectors. These are checked as in the program for jacobi. That is, the original matrix c is applied to each eigenvector, and the result is divided (element by element) by the eigenvector. Look for a result equal to the eigenvalue. (Note: in some cases, the vector element is zero or nearly so. These cases are flagged with the words "div. by zero".)

```
      PROGRAM xtqli
C     driver for routine tqli
      INTEGER NP
      REAL TINY
      PARAMETER(NP=10,TINY=1.0e-6)
      INTEGER i,j,k
      REAL a(NP,NP),c(NP,NP),d(NP),e(NP),f(NP)
      DATA c/5.0,4.3,3.0,2.0,1.0,0.0,-1.0,-2.0,-3.0,-4.0,
     *    4.3,5.1,4.0,3.0,2.0,1.0,0.0,-1.0,-2.0,-3.0,
     *    3.0,4.0,5.0,4.0,3.0,2.0,1.0,0.0,-1.0,-2.0,
     *    2.0,3.0,4.0,5.0,4.0,3.0,2.0,1.0,0.0,-1.0,
     *    1.0,2.0,3.0,4.0,5.0,4.0,3.0,2.0,1.0,0.0,
     *    0.0,1.0,2.0,3.0,4.0,5.0,4.0,3.0,2.0,1.0,
     *    -1.0,0.0,1.0,2.0,3.0,4.0,5.0,4.0,3.0,2.0,
     *    -2.0,-1.0,0.0,1.0,2.0,3.0,4.0,5.0,4.0,3.0,
     *    -3.0,-2.0,-1.0,0.0,1.0,2.0,3.0,4.0,5.0,4.0,
     *    -4.0,-3.0,-2.0,-1.0,0.0,1.0,2.0,3.0,4.0,5.0/
      do 12 i=1,NP
        do 11 j=1,NP
          a(i,j)=c(i,j)
11      continue
12    continue
      call tred2(a,NP,NP,d,e)
      call tqli(d,e,NP,NP,a)
      write(*,'(/1x,a)') 'Eigenvectors for a real symmetric matrix'
      do 16 i=1,NP
        do 14 j=1,NP
          f(j)=0.0
          do 13 k=1,NP
            f(j)=f(j)+c(j,k)*a(k,i)
13        continue
14      continue
        write(*,'(/1x,a,i3,a,f10.6)') 'Eigenvalue',i,' =',d(i)
        write(*,'(/1x,t7,a,t17,a,t31,a)') 'Vector','Mtrx*Vect.','Ratio'
        do 15 j=1,NP
          if (abs(a(j,i)).lt.TINY) then
            write(*,'(1x,2f12.6,a12)') a(j,i),f(j),'div. by 0'
          else
            write(*,'(1x,2f12.6,e14.6)') a(j,i),f(j),
```

```
      *               f(j)/a(j,i)
            endif
15          continue
            write(*,'(/1x,a)') 'press ENTER to continue...'
            read(*,*)
16          continue
            END
```

balanc reduces error in eigenvalue problems involving non-symmetric matrices.
It does this by adjusting corresponding rows and columns to have comparable norms,
without changing eigenvalues. Sample program xbalanc prepares the following array
a for balanc

$$\begin{pmatrix} 1 & 100 & 1 & 100 & 1 \\ 1 & 1 & 1 & 1 & 1 \\ 1 & 100 & 1 & 100 & 1 \\ 1 & 1 & 1 & 1 & 1 \\ 1 & 100 & 1 & 100 & 1 \end{pmatrix}$$

The norms of the five rows and five columns are printed out. It is clear from the
array that three of the rows and two of the columns have much larger norms than
the others. After balancing with balanc, the norms are recalculated, and this time
the row, column pairs should be much more nearly equal.

```
            PROGRAM xbalanc
C           driver for routine balanc
            INTEGER NP
            PARAMETER(NP=5)
            INTEGER i,j
            REAL a(NP,NP),r(NP),c(NP)
            DATA a/1.0,1.0,1.0,1.0,1.0,100.0,1.0,100.0,1.0,100.0,
      *         1.0,1.0,1.0,1.0,1.0,100.0,1.0,100.0,1.0,100.0,
      *         1.0,1.0,1.0,1.0,1.0/
C           print norms
            do 12 i=1,NP
              r(i)=0.0
              c(i)=0.0
              do 11 j=1,NP
                r(i)=r(i)+abs(a(i,j))
                c(i)=c(i)+abs(a(j,i))
11            continue
12          continue
            write(*,*) 'Rows:'
            write(*,*) (r(i),i=1,NP)
            write(*,*) 'Columns:'
            write(*,*) (c(i),i=1,NP)
            write(*,'(/1x,a/)') '***** Balancing Matrix *****'
            call balanc(a,NP,NP)
C           print norms
            do 14 i=1,NP
              r(i)=0.0
              c(i)=0.0
              do 13 j=1,NP
                r(i)=r(i)+abs(a(i,j))
                c(i)=c(i)+abs(a(j,i))
13            continue
```

```
14      continue
        write(*,*) 'Rows:'
        write(*,*) (r(i),i=1,NP)
        write(*,*) 'Columns:'
        write(*,*) (c(i),i=1,NP)
        END
```

elmhes reduces a general matrix to Hessenberg form using Gaussian elimination.
It is particularly valuable for real, non-symmetric matrices. Sample program xelmhes
employs balanc and elmhes to get a non-symmetric and grossly unbalanced matrix
into Hessenberg form. The matrix a is

$$\begin{pmatrix} 1 & 2 & 300 & 4 & 5 \\ 2 & 3 & 400 & 5 & 6 \\ 3 & 4 & 5 & 6 & 7 \\ 4 & 5 & 600 & 7 & 8 \\ 5 & 6 & 700 & 8 & 9 \end{pmatrix}$$

After printing the original matrix, the program feeds it to balanc and prints the
balanced version. This is submitted to elmhes and the result is printed. Notice that
the elements of a with $i>j+1$ are all set to zero by the program, because elmhes
returns random values in this part of the matrix. Therefore, the output of the test
program is guaranteed to *look* Hessenberg. To check that it is the *correct* Hessenberg
matrix, one has to go further and compute its eigenvalues, as will be done in the
example program after this one. We include here the expected results for comparison.

Balanced Matrix:

1.00	2.00	37.50	4.00	5.00
2.00	3.00	50.00	5.00	6.00
24.00	32.00	5.00	48.00	56.00
4.00	5.00	75.00	7.00	8.00
5.00	6.00	87.50	8.00	9.00

Reduced to Hessenberg Form:

.1000E+01	.3938E+02	.9618E+01	.3333E+01	.4000E+01
.2400E+02	.2733E+02	.1161E+03	.4800E+02	.4800E+02
.0000E+00	.8551E+02	-.4780E+01	-.1333E+01	-.2000E+01
.0000E+00	.0000E+00	.5188E+01	.1447E+01	.2171E+01
.0000E+00	.0000E+00	.0000E+00	-.9155E-07	.7874E-07

```
        PROGRAM xelmhes
C       driver for routine elmhes
        INTEGER NP
        PARAMETER(NP=5)
        INTEGER i,j
        REAL a(NP,NP)
        DATA a/1.0,2.0,3.0,4.0,5.0,2.0,3.0,4.0,5.0,6.0,
     *      300.0,400.0,5.0,600.0,700.0,4.0,5.0,6.0,7.0,8.0,
     *      5.0,6.0,7.0,8.0,9.0/
        write(*,'(/1x,a/)') '***** Original Matrix *****'
        do 11 i=1,NP
          write(*,'(1x,5f12.2)') (a(i,j),j=1,NP)
11      continue
        write(*,'(/1x,a/)') '***** Balance Matrix *****'
        call balanc(a,NP,NP)
```

```
      do 12 i=1,NP
        write(*,'(1x,5f12.2)') (a(i,j),j=1,NP)
12    continue
      write(*,'(/1x,a/)') '***** Reduce to Hessenberg Form *****'
      call elmhes(a,NP,NP)
      do 14 j=1,NP-2
        do 13 i=j+2,NP
          a(i,j)=0.0
13      continue
14    continue
      do 15 i=1,NP
        write(*,'(1x,5e12.4)') (a(i,j),j=1,NP)
15    continue
      END
```

hqr, finally, is a routine for finding the eigenvalues of a Hessenberg matrix using the QR algorithm. The 5×5 matrix a specified in the DATA statement is treated just as you would expect to treat any general real non-symmetric matrix. It is fed to balanc for balancing, to elmhes for reduction to Hessenberg form, and to hqr for eigenvalue determination. The eigenvalues may be complex-valued, and both real and imaginary parts are given. The original matrix has enough strategically placed zeros in it that you should have no trouble finding the eigenvalues by hand. Alternatively, you may check them against the list below:

Matrix:

1.00	2.00	.00	.00	.00
-2.00	3.00	.00	.00	.00
3.00	4.00	50.00	.00	.00
-4.00	5.00	-60.00	7.00	.00
-5.00	6.00	-70.00	8.00	-9.00

Eigenvalues:

#	Real	Imag.
1	.500000E+02	.000000E+00
2	.200000E+01	-.173205E+01
3	.200000E+01	.173205E+01
4	.700000E+01	.000000E+00
5	-.900000E+01	.000000E+00

```
      PROGRAM xhqr
C     driver for routine hqr
      INTEGER NP
      PARAMETER(NP=5)
      INTEGER i,j
      REAL a(NP,NP),wr(NP),wi(NP)
      DATA a/1.0,-2.0,3.0,-4.0,-5.0,2.0,3.0,4.0,5.0,6.0,
     *    0.0,0.0,50.0,-60.0,-70.0,0.0,0.0,0.0,7.0,8.0,
     *    0.0,0.0,0.0,0.0,0.0,-9.0/
      write(*,'(/1x,a)') 'Matrix:'
      do 11 i=1,NP
        write(*,'(1x,5f12.2)') (a(i,j),j=1,NP)
11    continue
      call balanc(a,NP,NP)
      call elmhes(a,NP,NP)
      call hqr(a,NP,NP,wr,wi)
```

```
      write(*,'(/1x,a)') 'Eigenvalues:'
      write(*,'(/1x,t9,a,t24,a/)') 'Real','Imag.'
      do 12 i=1,NP
        write(*,'(1x,2e15.6)') wr(i),wi(i)
12    continue
      END
```

Chapter 12: Fast Fourier Transform

Chapter 12 of Numerical Recipes deals with the fast Fourier transform (FFT). Routine `four1` *performs the FFT on a complex data array.* `twofft` *does the same transform on two real-valued data arrays (at the same time) and returns two complex-valued transforms. Finally,* `realft` *finds the Fourier transform of a single real-valued array. Two related transforms are the sine transform and the cosine transform, which comes in two incarnations. The corresponding routines are* `sinft`, `cosft1` *and* `cosft2`. *For FFTs in two or more dimensions the routine* `fourn` *is supplied. Real data in two or three dimensions is more efficiently handled by* `rlft3`, *while huge data sets on external media can be FFT'd by* `fourfs`.

$$\star \quad \star \quad \star \quad \star$$

Routine `four1` performs the fast Fourier transform on a complex-valued array of data points. Example program `xfour1` has five tests for this transform. First, it checks the following four symmetries (where $h(t)$ is the data and $H(n)$ is the transform):

1. If $h(t)$ is real-valued and even, then $H(n) = H(N - n)$ and H is real.

2. If $h(t)$ is imaginary-valued and even, then $H(n) = H(N-n)$ and H is imaginary.

3. If $h(t)$ is real-valued and odd, then $H(n) = -H(N - n)$ and H is imaginary.

4. If $h(t)$ is imaginary-valued and odd, then $H(n) = -H(N - n)$ and H is real.

The fifth test is that if a data array is Fourier transformed twice in succession, the resulting array should be identical to the original.

```
      PROGRAM xfour1
C     driver for routine four1
      INTEGER NN,NN2
      PARAMETER (NN=32,NN2=2*NN)
      REAL data(NN2),dcmp(NN2)
      INTEGER i,isign,j
      write(*,*) 'h(t)=real-valued even-function'
      write(*,*) 'H(n)=H(N-n) and real?'
      do 11 i=1,2*NN-1,2
        data(i)=1.0/((((i-NN-1.0)/NN)**2+1.0)
        data(i+1)=0.0
11    continue
      isign=1
      call four1(data,NN,isign)
      call prntft(data,NN2)
      write(*,*) 'h(t)=imaginary-valued even-function'
      write(*,*) 'H(n)=H(N-n) and imaginary?'
```

```
      do 12 i=1,2*NN-1,2
        data(i+1)=1.0/(((i-NN-1.0)/NN)**2+1.0)
        data(i)=0.0
12    continue
      isign=1
      call four1(data,NN,isign)
      call prntft(data,NN2)
      write(*,*) 'h(t)=real-valued odd-function'
      write(*,*) 'H(n)=-H(N-n) and imaginary?'
      do 13 i=1,2*NN-1,2
        data(i)=(i-NN-1.0)/NN/(((i-NN-1.0)/NN)**2+1.0)
        data(i+1)=0.0
13    continue
      data(1)=0.0
      isign=1
      call four1(data,NN,isign)
      call prntft(data,NN2)
      write(*,*) 'h(t)=imaginary-valued odd-function'
      write(*,*) 'H(n)=-H(N-n) and real?'
      do 14 i=1,2*NN-1,2
        data(i+1)=(i-NN-1.0)/NN/(((i-NN-1.0)/NN)**2+1.0)
        data(i)=0.0
14    continue
      data(2)=0.0
      isign=1
      call four1(data,NN,isign)
      call prntft(data,NN2)
C     transform, inverse-transform test
      do 15 i=1,2*NN-1,2
        data(i)=1.0/((0.5*(i-NN-1)/NN)**2+1.0)
        dcmp(i)=data(i)
        data(i+1)=(0.25*(i-NN-1)/NN)*
     *        exp(-(0.5*(i-NN-1.0)/NN)**2)
        dcmp(i+1)=data(i+1)
15    continue
      isign=1
      call four1(data,NN,isign)
      isign=-1
      call four1(data,NN,isign)
      write(*,'(/1x,t10,a,t44,a)') 'Original Data:',
     *     'Double Fourier Transform:'
      write(*,'(/1x,t5,a,t11,a,t24,a,t41,a,t53,a/)')
     *     'k','Real h(k)','Imag h(k)','Real h(k)','Imag h(k)'
      do 16 i=1,NN,2
        j=(i+1)/2
        write(*,'(1x,i4,2x,2f12.6,5x,2f12.6)') j,dcmp(i),
     *       dcmp(i+1),data(i)/NN,data(i+1)/NN
16    continue
      END

      SUBROUTINE prntft(data,nn2)
      INTEGER n,nn2,m,mm
      REAL data(nn2)
      write(*,'(/1x,t5,a,t11,a,t23,a,t39,a,t52,a)')
     *     'n','Real H(n)','Imag H(n)','Real H(N-n)','Imag H(N-n)'
      write(*,'(1x,i4,2x,2f12.6,5x,2f12.6)') 0,data(1),data(2),
     *     data(1),data(2)
```

```
      do 11 n=3,(nn2/2)+1,2
        m=(n-1)/2
        mm=nn2+2-n
        write(*,'(1x,i4,2x,2f12.6,5x,2f12.6)') m,data(n),
     *       data(n+1),data(mm),data(mm+1)
11    continue
      write(*,'(/1x,a)') ' press RETURN to continue ...'
      read(*,*)
      return
      END
```

twofft is a routine that performs an efficient FFT of two real arrays at once by packing them into a complex array and transforming with four1. Sample program xtwofft generates two periodic data sets, out of phase with one another, and performs a transform and an inverse transform on each. It will be difficult to judge whether the transform itself gives the right answer, but if the inverse transform gets you back to the easily recognized original, you may be fairly confident that the routine works.

```
      PROGRAM xtwofft
C     driver for routine twofft
      INTEGER N,N2
      REAL PER,PI
      PARAMETER(N=32,N2=2*N,PER=8.0,PI=3.14159)
      INTEGER i,isign
      REAL data1(N),data2(N),fft1(N2),fft2(N2),x
      do 11 i=1,N
        x=2.0*PI*i/PER
        data1(i)=nint(cos(x))
        data2(i)=nint(sin(x))
11    continue
      call twofft(data1,data2,fft1,fft2,N)
      write(*,*) 'Fourier transform of first function:'
      call prntft(fft1,N2)
      write(*,*) 'Fourier transform of second function:'
      call prntft(fft2,N2)
C     invert transform
      isign=-1
      call four1(fft1,N,isign)
      write(*,*) 'Inverted transform = first function:'
      call prntft(fft1,N2)
      call four1(fft2,N,isign)
      write(*,*) 'Inverted transform = second function:'
      call prntft(fft2,N2)
      END

      SUBROUTINE prntft(data,n2)
      INTEGER i,n2,nn2,m
      REAL data(n2)
      write(*,'(1x,t7,a,t13,a,t24,a,t35,a,t47,a)')
     *     'n','Real(n)','Imag.(n)','Real(N-n)','Imag.(N-n)'
      write(*,'(1x,i6,4f12.6)') 0,data(1),data(2),data(1),data(2)
      do 11 i=3,(n2/2)+1,2
        m=(i-1)/2
        nn2=n2+2-i
        write(*,'(1x,i6,4f12.6)') m,data(i),data(i+1),
     *       data(nn2),data(nn2+1)
```

```
11      continue
        write(*,'(/1x,a)') ' press RETURN to continue ...'
        read(*,*)
        return
        END
```

realft performs the Fourier transform of a single real-valued data array. Sample routine xrealft takes this function to be sinusoidal, and allows you to choose the period. After transforming, it simply plots the magnitude of each element of the transform. If the period you choose is a power of two, the transform will be nonzero in a single bin; otherwise there will be leakage to adjacent channels. xrealft follows every transform by an inverse transform to make sure the original function is recovered.

```
        PROGRAM xrealft
C       driver for routine realft
        INTEGER NP
        REAL EPS,PI,WIDTH
        PARAMETER(EPS=1.0e-3,NP=32,WIDTH=50.0,PI=3.14159)
        INTEGER i,j,n,nlim
        REAL big,per,scal,small,data(NP),size(NP)
        n=NP/2
1       write(*,'(1x,a,i2,a)') 'Period of sinusoid in channels (2-',
     *      NP,', OR 0 TO STOP)'
        read(*,*) per
        if (per.le.0.) stop
        do 11 i=1,NP
          data(i)=cos(2.0*PI*(i-1)/per)
11      continue
        call realft(data,NP,+1)
        size(1)=data(1)
        big=size(1)
        do 12 i=2,n
          size(i)=sqrt(data(2*i-1)**2+data(2*i)**2)
          if (size(i).gt.big) big=size(i)
12      continue
        scal=WIDTH/big
        do 13 i=1,n
          nlim=scal*size(i)+EPS
          write(*,'(1x,i4,1x,60a1)') i,('*',j=1,nlim+1)
13      continue
        write(*,*) 'press continue ...'
        read(*,*)
        call realft(data,NP,-1)
        big=-1.0e10
        small=1.0e10
        do 14 i=1,NP
          if (data(i).lt.small) small=data(i)
          if (data(i).gt.big) big=data(i)
14      continue
        scal=WIDTH/(big-small)
        do 15 i=1,NP
          nlim=scal*(data(i)-small)+EPS
          write(*,'(1x,i4,1x,60a1)') i,('*',j=1,nlim+1)
15      continue
        goto 1
```

```
        END
```

sinft performs a sine-transform of a real-valued array. The necessity for such a transform arises in solution methods for partial differential equations with certain kinds of boundary conditions (see Chapter 19). The sample program xsinft works exactly as the previous program. Notice that in this program no distinction needs to be made between the transform and its inverse. They are identical.

```
        PROGRAM xsinft
C       driver for routine sinft
        INTEGER NP
        REAL EPS,PI,WIDTH
        PARAMETER(EPS=1.0e-3,NP=16,WIDTH=30.0,PI=3.14159)
        INTEGER i,j,nlim
        REAL big,per,scal,small,data(NP)
1       write(*,'(1x,a,i2,a)') 'Period of sinusoid in channels (2-',NP,')'
        read(*,*) per
        if (per.le.0.) stop
        do 11 i=1,NP
          data(i)=sin(2.0*PI*(i-1)/per)
11      continue
        call sinft(data,NP)
        big=-1.0e10
        small=1.0e10
        do 12 i=1,NP
          if (data(i).lt.small) small=data(i)
          if (data(i).gt.big) big=data(i)
12      continue
        scal=WIDTH/(big-small)
        do 13 i=1,NP
          nlim=scal*(data(i)-small)+EPS
          write(*,'(1x,i4,1x,60a1)') i,('*',j=1,nlim+1)
13      continue
        write(*,*) 'press continue ...'
        read(*,*)
        call sinft(data,NP)
        big=-1.0e10
        small=1.0e10
        do 14 i=1,NP
          if (data(i).lt.small) small=data(i)
          if (data(i).gt.big) big=data(i)
14      continue
        scal=WIDTH/(big-small)
        do 15 i=1,NP
          nlim=scal*(data(i)-small)+EPS
          write(*,'(1x,i4,1x,60a1)') i,('*',j=1,nlim+1)
15      continue
        goto 1
        END
```

cosft1 is a companion subroutine to sinft that does the cosine transform. It also plays a role in partial differential equation solutions. Program xcosft1 is again the same as xrealft.

```
        PROGRAM xcosft1
C       driver for routine cosft1
        INTEGER NP
```

```
        REAL EPS,PI,WIDTH
        PARAMETER(EPS=1.0E-3,NP=17,WIDTH=30.0,PI=3.14159)
        INTEGER i,j,nlim
        REAL big,per,scal,small,data(NP)
1       write(*,'(1x,a,i2,a)') 'Period of cosine in channels (2-',NP,')'
        read(*,*) per
        if (per.le.0.) stop
        do 11 i=1,NP
          data(i)=cos(2.0*PI*(i-1)/per)
11      continue
        call cosft1(data,NP-1)
        big=-1.0e10
        small=1.0e10
        do 12 i=1,NP
          if (data(i).lt.small) small=data(i)
          if (data(i).gt.big) big=data(i)
12      continue
        scal=WIDTH/(big-small)
        do 13 i=1,NP
          nlim=scal*(data(i)-small)+EPS
          write(*,'(1x,i2,f6.2,1x,60a1)') i,data(i),('*',j=1,nlim+1)
13      continue
        write(*,*) 'press continue ...'
        read(*,*)
        call cosft1(data,NP-1)
        big=-1.0e10
        small=1.0e10
        do 14 i=1,NP
          if (data(i).lt.small) small=data(i)
          if (data(i).gt.big) big=data(i)
14      continue
        scal=WIDTH/(big-small)
        do 15 i=1,NP
          nlim=scal*(data(i)-small)+EPS
          write(*,'(1x,i4,1x,60a1)') i,('*',j=1,nlim+1)
15      continue
        goto 1
        END
```

cosft2 carries out a cosine transform on a "staggered" grid. Its demonstration program, xcosft2, is again the same as xrealft, except that the data array is filled on the staggered grid points.

```
        PROGRAM xcosft2
C       driver for routine cosft2
        INTEGER NP
        REAL EPS,PI,WIDTH
        PARAMETER(EPS=1.0E-3,NP=16,WIDTH=30.0,PI=3.14159)
        INTEGER i,j,nlim
        REAL big,per,scal,small,data(NP)
1       write(*,'(1x,a,i2,a)') 'Period of cosine in channels (2-',NP,')'
        read(*,*) per
        if (per.le.0.) stop
        do 11 i=1,NP
          data(i)=cos(2.0*PI*(i-0.5)/per)
11      continue
        call cosft2(data,NP,+1)
```

```
      big=-1.0e10
      small=1.0e10
      do 12 i=1,NP
        if (data(i).lt.small) small=data(i)
        if (data(i).gt.big) big=data(i)
12    continue
      scal=WIDTH/(big-small)
      do 13 i=1,NP
        nlim=scal*(data(i)-small)+EPS
        write(*,'(1x,i2,f6.2,1x,60a1)') i,data(i),('*',j=1,nlim+1)
13    continue
      write(*,*) 'press continue ...'
      read(*,*)
      call cosft2(data,NP,-1)
      big=-1.0e10
      small=1.0e10
      do 14 i=1,NP
        if (data(i).lt.small) small=data(i)
        if (data(i).gt.big) big=data(i)
14    continue
      scal=WIDTH/(big-small)
      do 15 i=1,NP
        nlim=scal*(data(i)-small)+EPS
        write(*,'(1x,i4,1x,60a1)') i,('*',j=1,nlim+1)
15    continue
      goto 1
      END
```

fourn is a routine for performing N-dimensional Fourier transforms. We have used it in sample program xfourn to transform a 3-dimensional complex data array of dimensions $4 \times 8 \times 16$. The function analyzed is simply an array of random values. The test conducted here is to perform a 3-dimensional transform and inverse transform in succession, and to compare the result with the original array. Ratios are provided for convenience. In the program, the 3-dimensional complex array is stored in a 1-dimensional real array data2 as described in *Numerical Recipes*. The array data1 is a saved copy of data2 for the comparison at the end.

```
      PROGRAM xfourn
C     driver for routine fourn
      INTEGER NDAT,NDIM
      PARAMETER(NDIM=3,NDAT=1024)
      INTEGER i,idum,isign,j,k,l,nn(NDIM)
      REAL data1(NDAT),data2(NDAT),ran1
      idum=-23
      do 11 i=1,NDIM
        nn(i)=2*(2**i)
11    continue
      do 14 i=1,nn(3)
        do 13 j=1,nn(2)
          do 12 k=1,nn(1)
            l=k+(j-1)*nn(1)+(i-1)*nn(2)*nn(1)
            l=2*l-1
C     real part of component
            data1(l)=2.*ran1(idum)-1.
            data2(l)=data1(l)
C     imaginary part of component
```

```
             l=l+1
             data1(l)=2.*ran1(idum)-1.
             data2(l)=data1(l)
12        continue
13        continue
14     continue
       isign=+1
       call fourn(data2,nn,NDIM,isign)
C      here would be any processing to be done in Fourier space
       isign=-1
       call fourn(data2,nn,NDIM,isign)
       write(*,'(1x,a)') 'Double 3-dimensional Transform'
       write(*,'(/1x,t10,a,t35,a,t63,a)') 'Double Transf.',
     *      'Original Data','Ratio'
       write(*,'(1x,t8,a,t20,a,t33,a,t45,a,t57,a,t69,a/)')
     *      'Real','Imag.','Real','Imag.','Real','Imag.'
       do 15 i=1,4
         j=2*i
         k=2*j
         l=k+(j-1)*nn(1)+(i-1)*nn(2)*nn(1)
         l=2*l-1
         write(*,'(1x,6f12.2)') data2(l),data2(l+1),data1(l),
     *        data1(l+1),data2(l)/data1(l),data2(l+1)/data1(l+1)
15        continue
       write(*,'(/1x,a,i4)') 'The product of transform lengths is:',
     *      nn(1)*nn(2)*nn(3)
       END
```

When your multidimensional data consists of real, rather than complex, values, it is faster (and more efficient in memory use) to use an FFT routine that is optimized for real data. rlft3 is such a routine. The main book already gives, at least in outline, three sample programs using this routine. Here we content ourselves with an example that simply fills a three dimensional real array with random values, performs the Fourier transform followed by its inverse, and then verifies that (up to a known constant multiplicative constant) the original matrix is recovered.

The function icompare is a simple utility for comparing all the values of an array for equality with another array to within a certain tolerance (allowing for roundoff errors). If the tolerance 0.0008 looks large to you, it is because we are comparing numbers whose magnitude is as large as $\sim 2 \times 10^3$, so the fractional tolerance is actually parts in 10^{-7}.

```
       PROGRAM xrlft3
C      driver for routine rlft3
       INTEGER NX,NY,NZ
       REAL EPS
       PARAMETER(NX=16,NY=8,NZ=32,EPS=0.0008)
       INTEGER icompare,idum,i,j,k,ierr
       REAL data1(NX,NY,NZ),data2(NX,NY,NZ),fnorm,ran1
       COMPLEX speq1(NY,NZ)
       INTEGER nn1,nn2,nn3
       idum=-3
       nn1=NX
       nn2=NY
       nn3=NZ
       fnorm=float(nn1)*float(nn2)*float(nn3)/2.
```

```
      do 13 i=1,nn1
        do 12 j=1,nn2
          do 11 k=1,nn3
            data1(i,j,k)=2.*ran1(idum)-1.
            data2(i,j,k)=data1(i,j,k)*fnorm
11        continue
12      continue
13    continue
      call rlft3(data1,speq1,nn1,nn2,nn3,1)
C     here would be any processing in Fourier space
      call rlft3(data1,speq1,nn1,nn2,nn3,-1)
      ierr=icompare('data',data1,data2,nn1*nn2*nn3,EPS)
      if (ierr.eq.0) then
        write(*,*) 'Data compares OK to tolerance',EPS
      else
        write(*,*) 'Comparison errors occured at tolerance',EPS
        write(*,*) 'Total number of errors is',ierr
      endif
      END

      INTEGER FUNCTION icompare(string,arr1,arr2,len,eps)
      INTEGER IPRNT
      PARAMETER(IPRNT=20)
      CHARACTER*(*) string
      INTEGER j,len
      REAL eps,arr1(len),arr2(len)
      write(*,*) string
      icompare=0
      do 11 j=1,len
        if ((arr2(j).eq.0.and.abs(arr1(j)-arr2(j)).gt.eps).or.
     *    (abs((arr1(j)-arr2(j))/arr2(j)).gt.eps)) then
          icompare=icompare+1
          if (icompare.le.IPRNT) write(*,*) j,arr1(j),arr2(j)
        endif
11    continue
      return
      END
```

If your need to take the Fourier transform of a data set that is simply too long to hold in your computer's memory, but which can be read from an external medium such as a tape or disk, then the routine `fourfs` can be used. The following example program creates a "pretend" such data set corresponding to a complex, 3-dimensional array (it actually fits perfectly well into the arrays `data1` and `data2`), writes its first half to one sequential file and its second half to another. `fourfs` is then called to take the the Fourier transform of the sequential files. To demonstrate that the correct answer is obtained, we use the in-memory routine `fourn` to take the same transform, and compare the answers.

Note that, for multidimensional arrays, the output of `fourfs` is the transpose of that from `fourn`. This is taken into account when the comparison is made.

The example program then inverts the Fourier transform. Since we are starting with the transpose of the inverse, the components of the array `idim` must be reversed for `fourfs`, but restored for `fourn`. A final comparison of values is made.

```
      PROGRAM xfourfs
C     driver for routine fourfs
      INTEGER NX,NY,NZ,NDAT
      PARAMETER (NX=8,NY=32,NZ=4,NDAT=2*NX*NY*NZ)
      INTEGER idum,i,j,k,l,kbf,nnv,idim(3),iunit(4)
      REAL diff,ran1,smax,sum,sum1,sum2,tot,data1(NDAT),data2(NDAT)
      COMPLEX cdata1(NX,NY,NZ),cdata2(NX,NY,NZ),ctdat(NZ,NY,NX)
      EQUIVALENCE (data1,cdata1,ctdat),(data2,cdata2)
      kbf=128
      nnv=3
      idim(1)=NX
      idim(2)=NY
      idim(3)=NZ
      tot=float(NX)*float(NY)*float(NZ)
      open (unit=10,form='unformatted',status='scratch')
      open (unit=11,form='unformatted',status='scratch')
      open (unit=12,form='unformatted',status='scratch')
      open (unit=13,form='unformatted',status='scratch')
      do 11 i=1,4
        iunit(i)=i+9
11    continue
      idum=-23
      do 14 i=1,idim(3)
        do 13 j=1,idim(2)
          do 12 k=1,idim(1)
            l=k+(j-1)*idim(1)+(i-1)*idim(2)*idim(1)
            l=2*l-1
            data1(l)=2.*ran1(idum)-1.
            data2(l)=data1(l)
            l=l+1
            data1(l)=2.*ran1(idum)-1.
            data2(l)=data1(l)
12        continue
13      continue
14    continue
      do 15 j=1,NDAT/2,kbf
        write(iunit(1)) (data1(j+i),i=0,kbf-1)
        write(iunit(2)) (data1(NDAT/2+j+i),i=0,kbf-1)
15    continue
      write(*,*) '*************** now doing fourfs ********'
      call fourfs(iunit,idim,nnv,1)
      do 16 j=1,NDAT/2,kbf
        read(iunit(3)) (data1(j+i),i=0,kbf-1)
        read(iunit(4)) (data1(NDAT/2+j+i),i=0,kbf-1)
16    continue
      write(*,*) '*************** now doing fourn *********'
      call fourn(data2,idim,nnv,1)
      sum=0.
      smax=0.
      sum2=0.
      do 19 i=1,NZ
        do 18 j=1,NY
          do 17 k=1,NX
            diff=abs(cdata2(k,j,i)-ctdat(i,j,k))
            sum2=sum2+real(ctdat(i,j,k))**2+aimag(ctdat(i,j,k))**2
            sum=sum+diff
            if (diff.gt.smax) smax=diff
```

```
17          continue
18        continue
19      continue
        sum2=sqrt(sum2/tot)
        sum=sum/tot
        write(*,'(1x,a,3f12.7)')
     *    '(r.m.s.) value, (max,ave) discrepancy=',sum2,smax,sum
        do 21 i=1,4
          rewind(unit=iunit(i))
21      continue
C       now check the inverse transforms
        idim(1)=NZ
        idim(2)=NY
        idim(3)=NX
        idum=iunit(1)
        iunit(1)=iunit(3)
        iunit(3)=idum
        idum=iunit(2)
        iunit(2)=iunit(4)
        iunit(4)=idum
        write(*,*) '*************** now doing fourfs ********'
        call fourfs(iunit,idim,nnv,-1)
        do 22 j=1,NDAT/2,kbf
          read(iunit(3)) (data1(j+i),i=0,kbf-1)
          read(iunit(4)) (data1(NDAT/2+j+i),i=0,kbf-1)
22      continue
        idim(1)=NX
        idim(2)=NY
        idim(3)=NZ
        write(*,*) '*************** now doing fourn *********'
        call fourn(data2,idim,nnv,-1)
        sum=0.
        smax=0.
        sum1=0.
        do 25 i=1,NZ
          do 24 j=1,NY
            do 23 k=1,NX
              sum1=sum1+real(cdata2(k,j,i))**2+aimag(cdata2(k,j,i))**2
              diff=abs(cdata2(k,j,i)-cdata1(k,j,i))
              sum=sum+diff
              if (diff.gt.smax) smax=diff
23          continue
24        continue
25      continue
        sum=sum/tot
        sum1=sqrt(sum1/tot)
        write(*,'(1x,a,3f12.7)')
     *    '(r.m.s.) value, (max,ave) discrepancy=',sum1,smax,sum
        write(*,*) 'ratio of r.m.s. values, expected ratio=',
     *    sum1/sum2,sqrt(tot)
        END
```

Chapter 13: Fourier and Spectral Applications

Chapter 13 of Numerical Recipes covers mainly Fourier transform spectral methods, particularly the transform of discretely sampled data. Two common uses of the Fourier transform are the convolution of data with a response function, and the computation of the correlation of two data sets. These operations are carried out by convlv *and* correl *respectively. Other applications of Fourier methods in the main book include data filtering, power spectrum estimation (*spctrm, *or* evlmem *with* memcof*), and linear prediction (*predic *with* fixrts *and* memcof*). For power spectrum estimation of unevenly sampled data, you can use* period, *or its "fast" cousin for large data sets,* fasper. *An application with a somewhat different flavor is the high accuracy computation of Fourier integrals with* dftint. *Finally, we provide routines for wavelet transforms:* wt1 *for a one-dimensional wavelet transform,* wtn *for a multidimensional wavelet transform, and (in* daub4, pwtset, *and* pwt*) a small selection of wavelet filters, out of which the wavelets themselves are constructed.*

$\star$ $\quad$ $\star$ $\quad$ $\star$ $\quad$ $\star$

Subroutine convlv performs the convolution of a data set with a response function using an FFT. Sample program xconvlv uses two functions that take on only the values 0.0 and 1.0. The data array data(i) has sixteen values, and is zero everywhere except between i=6 and i=10 where it is 1.0. The response function respns(i) has nine values and is zero except between i=3 and i=6 where it is 1.0. The expected value of the convolution is determined simply by flipping the response function end-to-end, moving it to the left by the desired shift, and counting how many non-zero channels of respns fall on non-zero channels of data. In this way, you should be able to verify the result from the program. The sample program, incidentally, does the calculation by this direct method for the purpose of comparison.

```
      PROGRAM xconvlv
C     driver for routine convlv
      INTEGER N,N2,M
      REAL PI
      PARAMETER(N=16,N2=32,M=9,PI=3.14159265)
      INTEGER i,isign,j
      REAL cmp,data(N),respns(M),resp(N),ans(N2)
      do 11 i=1,N
        data(i)=0.0
        if ((i.ge.(N/2-N/8)).and.(i.le.(N/2+N/8))) data(i)=1.0
11    continue
      do 12 i=1,M
        respns(i)=0.0
```

```
      if (i.gt.2 .and. i.lt.7) respns(i)=1.0
      resp(i)=respns(i)
12    continue
      isign=1
      call convlv(data,N,resp,M,isign,ans)
C     compare with a direct convolution
      write(*,'(/1x,t4,a,t13,a,t24,a)') 'I','CONVLV','Expected'
      do 14 i=1,N
        cmp=0.0
        do 13 j=1,M/2
          cmp=cmp+data(mod(i-j-1+N,N)+1)*respns(j+1)
          cmp=cmp+data(mod(i+j-1,N)+1)*respns(M-j+1)
13      continue
        cmp=cmp+data(i)*respns(1)
        write(*,'(1x,i3,3x,2f12.6)') i,ans(i),cmp
14    continue
      END
```

correl calculates the correlation function of two data sets. Sample program xcorrel defines data1(i) as an array of 64 values that are all zero except from i=25 to i=39, where they are one. data2(i) is defined in the same way. Therefore, the correlation being performed is an autocorrelation. The sample routine compares the result of the calculation as performed by correl with that found by a direct calculation. In this case the calculation may be done manually simply by successively shifting data2 with respect to data1 and counting the number of nonzero channels of the two that overlap.

```
      PROGRAM xcorrel
C     driver for routine correl
      INTEGER N,N2
      REAL PI
      PARAMETER(N=64,N2=128,PI=3.1415927)
      INTEGER i,j
      REAL cmp,data1(N),data2(N),ans(N2)
      do 11 i=1,N
        data1(i)=0.0
        if ((i.gt.(N/2-N/8)).and.(i.lt.(N/2+N/8))) data1(i)=1.0
        data2(i)=data1(i)
11    continue
      call correl(data1,data2,N,ans)
C     calculate directly
      write(*,'(/1x,t4,a,t13,a,t25,a/)') 'n','CORREL','Direct Calc.'
      do 13 i=0,16
        cmp=0.0
        do 12 j=1,N
          cmp=cmp+data1(mod(i+j-1,N)+1)*data2(j)
12      continue
        write(*,'(1x,i3,3x,f12.6,f15.6)') i,ans(i+1),cmp
13    continue
      END
```

spctrm does a spectral estimate of a data set by reading it in as segments, windowing, Fourier transforming, and accumulating the power spectrum. Data segments may or may not be overlapped at the decision of the user. In sample program xspctrm the spectral data is read in from a file called SPCTRL.DAT containing 1200 numbers and included on the diskette. It is analyzed first with overlap and then without. The results are tabulated side by side for comparison.

```
      PROGRAM xspctrm
C     driver for routine spctrm
      INTEGER M,M4
      PARAMETER(M=16,M4=4*M)
      INTEGER j,k
      REAL p(M),q(M),w1(M4),w2(M)
      LOGICAL ovrlap
      open(9,file='SPCTRL.DAT',status='OLD')
      k=8
      ovrlap=.true.
      call spctrm(p,M,k,ovrlap,w1,w2)
      rewind(9)
      k=16
      ovrlap=.false.
      call spctrm(q,M,k,ovrlap,w1,w2)
      close(9)
      write(*,*) 'Spectrum of DATA in file SPCTRL.DAT'
      write(*,'(1x,t14,a,t29,a)') 'Overlapped','Non-Overlapped'
      do 11 j=1,M
        write(*,'(1x,i4,2f17.6)') j,p(j),q(j)
11    continue
      END
```

memcof is used in two different capacities in *Numerical Recipes*. In combination with evlmem it is used to perform spectral analysis by the maximum entropy method. memcof finds the coefficients for a model spectrum, the magnitude squared of the inverse of a polynomial series. Sample program xmemcof determines the coefficients for 1000 numbers from the file SPCTRL.DAT and simply prints the results for comparison to the following table:

```
Coefficients for spectral estimation of spctrl.dat
 a[ 1] =    1.261539
 a[ 2] =   -0.007695
 a[ 3] =   -0.646778
 a[ 4] =   -0.280603
 a[ 5] =    0.163693
 a[ 6] =    0.347674
 a[ 7] =    0.111247
 a[ 8] =   -0.337141
 a[ 9] =   -0.358043
 a[10] =    0.378774
    a0 =    0.003511
```

```
      PROGRAM xmemcof
C     driver for routine memcof
      INTEGER N,M
      PARAMETER(N=1000,M=10)
      INTEGER i
      REAL pm,data(N),cof(M)
      open(7,file='SPCTRL.DAT',status='OLD')
      read(7,*) (data(i),i=1,N)
      close(7)
      call memcof(data,N,M,pm,cof)
      write(*,'(/1x,a/)') 'Coeff. for spectral estim. of SPCTRL.DAT'
```

```
      do 11 i=1,M
        write(*,'(1x,a,i2,a,f12.6)')') 'a[',i,'] =',cof(i)
11    continue
      write(*,'(/1x,a,f12.6/)')') 'a0 =',pm
      END
```

Although `evlmem` appears later in Chapter 13, we show its sample program here, because it is so closely tied to the previous program. `evlmem` uses coefficients from `memcof` to generate a spectral estimate. The example `xevlmem` uses the data from `SPCTRL.DAT` and prints the spectral estimate. You may compare the result to:

```
Power spectrum estimate of DATA in SPCTRL.DAT
    f*delta        power
   0.000000     0.026023
   0.031250     0.029266
   0.062500     0.193087
   0.093750     0.139241
   0.125000    29.915518
   0.156250     0.003878
   0.187500     0.000633
   0.218750     0.000334
   0.250000     0.000437
   0.281250     0.001331
   0.312500     0.000780
   0.343750     0.000451
   0.375000     0.000784
   0.406250     0.001381
   0.437500     0.000649
   0.468750     0.000775
   0.500000     0.001716
```

```
      PROGRAM xevlmem
C     driver for routine evlmem
      INTEGER N,NFDT,M
      PARAMETER(N=1000,M=10,NFDT=16)
      INTEGER i
      REAL evlmem,fdt,pm,data(N),cof(M)
      open(7,file='SPCTRL.DAT',status='old')
      read(7,*) (data(i),i=1,N)
      close(7)
      call memcof(data,N,M,pm,cof)
      write(*,*) 'Power spectrum estimate of DATA in SPCTRL.DAT'
      write(*,'(1x,t6,a,t20,a)') 'f*delta','power'
      do 11 i=0,NFDT
        fdt=0.5*i/NFDT
        write(*,'(1x,2f12.6)') fdt,evlmem(fdt,cof,M,pm)
11    continue
      END
```

Notice that once `memcof` has determined coefficients, we may evaluate the estimate at any intervals we wish. Notice also that we have built a spectral peak into the noisy data in `SPCTRL.DAT`.

The second use of `memcof` is to perform linear prediction in combination with `predic` and `fixrts`. `memcof` produces the linear prediction coefficients from the data set. `fixrts` massages the coefficients so that all roots of the characteristic polynomial fall inside the unit circle of the complex domain, thus insuring stability of the prediction algorithm. Finally, `predic` predicts future data points based on the modified coefficients. Sample program `xfixrts` demonstrates the operation of `fixrts`. The coefficients provided in the `DATA` statement for `d(i)` are those appropriate to the polynomial $(z - 1)^6 = 1$. This equation has six roots on a circle of radius one, centered at $(1.0, 0.0)$ in the complex plane. Some of these lie within the unit circle and some outside. The ones outside are moved by `fixrts` according to $z_i \rightarrow 1/z_i{}^*$. You can easily figure these out by hand and check the results. Also, the sample routine calculates $(z - 1)^6$ for each of the adjusted roots, and thereby shows which have been changed and which have not.

```
      PROGRAM xfixrts
C     driver for routine fixrts
      INTEGER NPOL,NPOLES
      PARAMETER(NPOLES=6,NPOL=NPOLES+1)
      INTEGER i
      REAL d(NPOLES)
      COMPLEX zcoef(NPOL),zeros(NPOLES),z
      LOGICAL polish
      DATA d/6.0,-15.0,20.0,-15.0,6.0,0.0/
C     finding roots of (z-1.0)**6=1.0
C     first print roots
      zcoef(NPOLES+1)=cmplx(1.0,0.0)
      do 11 i=NPOLES,1,-1
        zcoef(i)=cmplx(-d(NPOLES+1-i),0.0)
11    continue
      polish=.true.
      call zroots(zcoef,NPOLES,zeros,polish)
      write(*,'(/1x,a)') 'Roots of (z-1.0)^6 = 1.0'
      write(*,'(1x,t20,a,t42,a)') 'Root','(z-1.0)^6'
      do 12 i=1,NPOLES
        z=(zeros(i)-1.0)**6
        write(*,'(1x,i6,4f12.6)') i,zeros(i),z
12    continue
C     now fix them to lie within unit circle
      call fixrts(d,NPOLES)
C     check results
      zcoef(NPOLES+1)=cmplx(1.0,0.0)
      do 13 i=NPOLES,1,-1
        zcoef(i)=cmplx(-d(NPOLES+1-i),0.0)
13    continue
      call zroots(zcoef,NPOLES,zeros,polish)
      write(*,'(/1x,a)') 'Roots reflected in unit circle'
      write(*,'(1x,t20,a,t42,a)') 'Root','(z-1.0)^6'
      do 14 i=1,NPOLES
        z=(zeros(i)-1.0)**6
        write(*,'(1x,i6,4f12.6)') i,zeros(i),z
14    continue
      END
```

`predic` carries out the job of performing the prediction. The function chosen for investigation in sample program `xpredic` is

$$F(n) = \exp(-n/npts)\sin(2\pi n/50) + \exp(-2n/npts)\sin(2.2\pi n/50)$$

the sum of two sine waves of similar period and exponentially decaying amplitudes. On the basis of 500 data points, and working with coefficients representing ten poles, the routine predicts 20 future points. The quality of this prediction may be judged by comparing these 20 points with the evaluations of $F(n)$ that are provided.

```
        PROGRAM xpredic
C       driver for routine predic
        INTEGER NPTS,NPOLES,NFUT
        REAL PI
        PARAMETER(NPTS=500,NPOLES=10,NFUT=20,PI=3.1415926)
        INTEGER i,n
        REAL f,dum,data(NPTS),d(NPOLES),future(NFUT)
        f(n)=exp(-1.0*n/NPTS)*sin(2.0*PI*n/50.0)
     *      +exp(-2.0*n/NPTS)*sin(2.2*PI*n/50.0)
        do 11 i=1,NPTS
          data(i)=f(i)
11      continue
        call memcof(data,NPTS,NPOLES,dum,d)
        call fixrts(d,NPOLES)
        call predic(data,NPTS,d,NPOLES,future,NFUT)
        write(*,'(6x,a,t13,a,t25,a)') 'I','Actual','PREDIC'
        do 12 i=1,NFUT
          write(*,'(1x,i6,2f12.6)') i,f(i+NPTS),future(i)
12      continue
        END
```

Spectral analysis of unevenly sampled data, or of evenly sampled data with missing values, can be performed by the routine `period`, which constructs the Lomb normalized periodogram. In the following example, we construct a data array of length 90 that consists of a regular sampling of length 100 with 10 (random) deletions. The sampled function is a pure cosine, plus Gaussian white noise of somewhat larger amplitude. A call to `period` yields the significance level with which the signal is found to be non-Gaussian (very significant in this case!), and the output spectrum. We print out a few values in the neighborhood of the largest peak found. You can verify that the peak is indeed nicely centered on the frequency of the input cosine.

```
        PROGRAM xperiod
C       driver for routine period
        INTEGER NP,NPR
        REAL TWOPI
        PARAMETER(NP=90,NPR=11,TWOPI=6.2831853)
        INTEGER idum,j,jmax,n,nout
        REAL gasdev,prob,x(NP),y(NP),px(2*NP),py(2*NP)
        idum=-4
        j=0
        do 11 n=1,NP+10
          if (n.ne.3.and.n.ne.4.and.n.ne.6.and.n.ne.21.and.
     *        n.ne.38.and.n.ne.51.and.n.ne.67.and.n.ne.68.and.
     *        n.ne.83.and.n.ne.93) then
            j=j+1
            x(j)=n
            y(j)=0.75*cos(0.6*x(j))+gasdev(idum)
          endif
11      continue
        call period(x,y,j,4.,1.,px,py,2*NP,nout,jmax,prob)
        write(*,*) 'PERIOD results for test signal (cos(0.6x) + noise):'
        write(*,*) 'NOUT,JMAX,PROB=',nout,jmax,prob
        do 12 n=max(1,jmax-NPR/2),min(nout,jmax+NPR/2)
```

```
      write(*,*) n,TWOPI*px(n),py(n)
12    continue
      END
```

The routine `fasper` has the same capabilities as `period`, but performs its underlying calculations using FFT routines. For large data arrays, `fasper` is very much faster than `period`. Here, we give a demonstration program that is closely equivalent to the one given for the previous routine. About the only difference is that `fasper` requires a larger workspace, here 4096, onto which it "extirpolates" the given data, as described in the main book.

```
      PROGRAM xfasper
C     driver for routine fasper
      INTEGER NP,MP,NPR
      REAL TWOPI
      PARAMETER(NP=90,MP=4096,NPR=11,TWOPI=6.2831853)
      INTEGER idum,j,jmax,n,nout
      REAL gasdev,prob,x(NP),y(NP),px(MP),py(MP)
      idum=-4
      j=0
      do 11 n=1,NP+10
        if (n.ne.3.and.n.ne.4.and.n.ne.6.and.n.ne.21.and.
     *        n.ne.38.and.n.ne.51.and.n.ne.67.and.n.ne.68.and.
     *        n.ne.83.and.n.ne.93) then
          j=j+1
          x(j)=n
          y(j)=0.75*cos(0.6*x(j))+gasdev(idum)
        endif
11    continue
      call fasper(x,y,j,4.,1.,px,py,MP,nout,jmax,prob)
      write(*,*) 'FASPER results for test signal (cos(0.6x) + noise):'
      write(*,*) 'NOUT,JMAX,PROB=',nout,jmax,prob
      do 12 n=max(1,jmax-NPR/2),min(nout,jmax+NPR/2)
        write(*,*) n,TWOPI*px(n),py(n)
12    continue
      END
```

We test `dftint` on the integrals

$$\int_a^b \cos(cx+d)\cos\omega x\,dx = \left[\frac{\sin[(\omega-c)x-d]}{2(\omega-c)} + \frac{\sin[(\omega+c)x+d]}{2(\omega+c)}\right]_a^b$$

$$\int_a^b \cos(cx+d)\sin\omega x\,dx = \left[-\frac{\cos[(\omega-c)x-d]}{2(\omega-c)} - \frac{\cos[(\omega+c)x+d]}{2(\omega+c)}\right]_a^b$$

where $\omega \neq \pm c$. Sample program `xdftint` prompts for values of c, d, a, b and ω. It then compares the results of `dftint` with the analytic values of the integrals.

```
      PROGRAM xdftint
C     driver for routine dftint
      REAL a,b,c,cans,cosint,d,sans,sinint,w
      COMMON /parms/ c,d
      EXTERNAL coscxd
      write(*,'(2x,a,t10,a,t37,a,t44,a,t70,a)') 'Omega',
     * 'Integral cosine*test func','Err','Integral sine*test func','Err'
3     write(*,*) 'INPUT C,D: '
```

```
      read(*,*) c,d
1     write(*,*) 'INPUT A,B: '
      read(*,*) a,b
      if (a.eq.b) goto 3
2     write(*,*) 'INPUT W: '
      read(*,*,END=999) w
      if (w.lt.0.) goto 1
      call dftint(coscxd,a,b,w,cosint,sinint)
      call getans(w,a,b,cans,sans)
      write(*,100) w,cans,cosint-cans,sans,sinint-sans
100   format(1p5e15.6)
      goto 2
999   write(*,*) 'NORMAL COMPLETION'
      STOP
      END


      REAL FUNCTION coscxd(x)
      REAL c,d,x
      COMMON /parms/ c,d
      coscxd=cos(c*x+d)
      return
      END


      SUBROUTINE getans(w,a,b,cans,sans)
      REAL a,b,c,cans,ci,d,sans,si,w,x
      COMMON /parms/ c,d
      ci(x)=sin((w-c)*x-d)/(2.*(w-c))+sin((w+c)*x+d)/(2.*(w+c))
      si(x)=-cos((w-c)*x-d)/(2.*(w-c))-cos((w+c)*x+d)/(2.*(w+c))
      cans=ci(b)-ci(a)
      sans=si(b)-si(a)
      return
      END
```

The wavelet transform routine wt1 can be used either with the lowest-order
wavelet filter, as provided by daub4, or else with the higher-order filters provided
by the combination of pwtset and pwt. The following sample program allows for
both possibilities by prompting for the input order as k. For k = −4 the program
uses daub4, while for k positive it uses pwtset(k). Also prompted for is a quantity
frac: The program computes the wavelet transform of a function consisting of a
parabolic "bump" on an otherwise zero background. It then truncates the wavelet
transform to zeros, except for a fraction frac of the components. After reconstructing
the function from the truncated data with an inverse wavelet transform, it finds the
maximum deviation between the original function and its approximate reconstruction. One finds that with a comparatively small number of wavelet components, one
is able to reconstruct the original bump to very good approximation.

```
      PROGRAM xwt1
C     driver for routine wt1
      INTEGER NCMAX,NMAX,NCEN,NWID
      PARAMETER(NMAX=512,NCMAX=50,NCEN=333,NWID=33)
      INTEGER i,itest,k,ncof,ioff,joff,nused
      REAL u(NMAX),v(NMAX),w(NMAX),cc,cr,frac,select,thresh,tmp
      COMMON /pwtcom/ cc(NCMAX),cr(NCMAX),ncof,ioff,joff
      EXTERNAL pwt,daub4
1     write(*,*) 'Enter k (4, -4, 12, or 20) and frac (0. to 1.):'
```

```
      read(*,*,END=999) k,frac
      frac=min(1.,max(0.,frac))
      if (k.eq.-4) then
        itest=1
      else
        itest=0
      endif
      k=abs(k)
      if (k.ne.4.and.k.ne.12.and.k.ne.20) goto 1
      do 11 i=1,NMAX
        if (i.gt.NCEN-NWID.and.i.lt.NCEN+NWID) then
          v(i)=float(i-NCEN+NWID)*float(NCEN+NWID-i)/NWID**2
        else
          v(i)=0.
        endif
        w(i)=v(i)
11    continue
      if (itest.eq.0) then
        call pwtset(k)
        call wt1(v,NMAX,1,pwt)
      else
        call wt1(v,NMAX,1,daub4)
      endif
      do 12 i=1,NMAX
        u(i)=abs(v(i))
12    continue
      thresh=select(int((1.-frac)*NMAX),NMAX,u)
      nused=0
      do 13 i=1,NMAX
        if (abs(v(i)).le.thresh) then
          v(i)=0.
        else
          nused=nused+1
        endif
13    continue
      if (itest.eq.0) then
        call wt1(v,NMAX,-1,pwt)
      else
        call wt1(v,NMAX,-1,daub4)
      endif
      thresh=0.
      do 14 i=1,NMAX
        tmp=abs(v(i)-w(i))
        if (tmp.gt.thresh) thresh=tmp
14    continue
      write(*,*) 'k,NMAX,nused=',k,NMAX,nused
      write(*,*) 'discrepancy=',thresh
      goto 1
999   END
```

As an example of using `wtn`, the n-dimensional wavelet transform routine, we set up the two-dimensional matrix described in the caption to Figure 13.10.5 of the main book, then take its wavelet transform (using the DAUB12 wavelet), then undo the transform and show that the original matrix is recovered. In real applications, of course, what matters is what we have left out: processes that would be applied to the matrix while it is in wavelet space. The simplest such process would be truncation, encoding, and transmission of the matrix in a wavelet-compressed form, for reconstruction at another location. Another kind of process would be truncation followed by some operations of linear algebra optimized for the now-sparse form.

```
      PROGRAM xwtn
C     driver for routine wtn
      INTEGER NCMAX,NX,NY
      REAL EPS
      PARAMETER(NCMAX=50,NX=128,NY=256,EPS=1.e-06)
      INTEGER ncof,ioff,joff,i,j,nerror,ntot
      REAL a(NX,NY),aorg(NX,NY),cc,cr
      COMMON /pwtcom/ cc(NCMAX),cr(NCMAX),ncof,ioff,joff
      INTEGER ndim(2)
      EXTERNAL pwt
      DATA ndim /NX,NY/
      nerror=0
      ntot=NX*NY
      do 12 i=1,NX
        do 11 j=1,NY
          if (i.eq.j) then
            a(i,j)=-1.
          else
            a(i,j)=1./sqrt(abs(float(i-j)))
          endif
          aorg(i,j)=a(i,j)
11      continue
12    continue
      call pwtset(12)
      call wtn(a,ndim,2,1,pwt)
C     here, one might set the smallest components to zero, encode and transmit
C     the remaining components as a compressed form of the "image"
      call wtn(a,ndim,2,-1,pwt)
      do 14 i=1,NX
        do 13 j=1,NY
          if (abs(aorg(i,j)-aorg(i,j)).ge.EPS) then
            write(*,*) 'Compare Error at element ',i,j
            nerror=nerror+1
          endif
13      continue
14    continue
      if (nerror.ne.0) then
        write(*,*) 'Number of comparision errors: ',nerror
      else
        write(*,*) 'Transform-inverse transform check OK'
      endif
      END
```

Chapter 14: Statistical Description of Data

Chapter 14 of *Numerical Recipes* covers the subject of descriptive statistics, the representation of data in terms of its statistical properties, and the use of such properties to compare data sets. We start with a subroutine to characterize data sets: `moment` returns the average, average deviation, standard deviation, variance, skewness, and kurtosis of a data array.

Most of the remaining subroutines compare data sets. `ttest` compares the means of two data sets having the same variance; `tutest` does the same for two sets having different variance; and `tptest` does it for paired samples, correcting for covariance. `ftest` is a test of whether two data arrays have significantly different variance. The question of whether two distributions are different is treated by four subroutines (pertaining to whether the data is binned or continuous, and whether data is compared to a model distribution or to other data). Specifically,

1. `chsone` compares binned data to a model distribution.

2. `chstwo` compares two binned data sets.

3. `ksone` compares the cumulative distribution function of an unbinned data set to a given function.

4. `kstwo` compares the cumulative distribution functions of two unbinned data sets.

The next set of subroutines tests for associations between nominal variables. `cntab1` and `cntab2` both check for associations in a two-dimensional contingency table, the first calculating on the basis of χ^2, and the second by evaluating entropies. Linear correlation is represented by Pearson's r, or the linear correlation coefficient, which is calculated with routine `pearsn`. Alternatively, the data can be investigated with a nonparametric or rank correlation, using `spear` to find Spearman's rank correlation r_s. Kendall's τ uses rank ordering of ordinal data to test for monotonic correlations. `kendl1` does this for two data arrays of the same size, while `kendl2` applies it to contingency tables.

We provide two subroutines for testing two-dimensional data sets. `ks2d1s` performs a Kolmogorov-Smirnov test of data to a given model (generalization of `ksone`), while `ks2d2s` does the same for testing one data set against another (generalization of `kstwo`).

One final routine `savgol` makes no attempt to describe or compare data statistically. It calculates Savitzky-Golay smoothing coefficients that can be used to smooth out statistical fluctuations, usually for the purpose of visual *presentation.*

⋆   ⋆   ⋆   ⋆

Subroutine `moment` calculates successive moments of a given distribution of data. The example program `xmoment` creates an unusual distribution, one that has a sinusoidal distribution of values (over a half-period of the sine, so the distribution is a symmetrical peak). We have worked out the moments of such a distribution theoretically and recorded them in the program for comparison. The data is discrete and will only approximate these values.

```
      PROGRAM xmoment
C     driver for routine moment
      REAL PI
      INTEGER NBIN,NDAT,NPTS
      PARAMETER(PI=3.14159265,NPTS=10000,NBIN=100,NDAT=NPTS+NBIN)
      INTEGER i,j,k,nlim
      REAL adev,ave,curt,data(NDAT),sdev,skew,var,x
      i=1
      do 12 j=1,NBIN
        x=PI*j/NBIN
        nlim=nint(sin(x)*PI/2.0*NPTS/NBIN)
        do 11 k=1,nlim
          data(i)=x
          i=i+1
11      continue
12    continue
      write(*,'(1x,a/)') 'Moments of a sinusoidal distribution'
      call moment(data,i-1,ave,adev,sdev,var,skew,curt)
      write(*,'(1x,t29,a,t42,a/)') 'Calculated','Expected'
      write(*,'(1x,a,t25,2f12.4)') 'Mean :',ave,PI/2.0
      write(*,'(1x,a,t25,2f12.4)') 'Average Deviation :',adev,0.570796
      write(*,'(1x,a,t25,2f12.4)') 'Standard Deviation :',sdev,0.683667
      write(*,'(1x,a,t25,2f12.4)') 'Variance :',var,0.467401
      write(*,'(1x,a,t25,2f12.4)') 'Skewness :',skew,0.0
      write(*,'(1x,a,t25,2f12.4)') 'Kurtosis :',curt,-0.806249
      END
```

Student's *t*-test is a test of two data sets for significantly different means. It is applied by `xttest` to two Gaussian data sets `data1` and `data2` that are generated by `gasdev`. `data2` is originally given an artificial shift of its mean to the right of that of `data1`, by `NSHFT/2` units of `EPS`. Then `data1` is successively shifted `NSHFT` times to the right by `EPS` and compared to `data2` by `ttest`. At about step `NSHFT/2`, the two distributions should superpose and indicate populations with the same mean. Notice that the two populations have the same variance (i.e. 1.0), as required by `ttest`.

```
      PROGRAM xttest
C     driver for routine ttest
      INTEGER NPTS,NSHFT,MPTS
      REAL EPS
      PARAMETER(NPTS=1024, MPTS=512, EPS=0.02,NSHFT=10)
      INTEGER i,idum,j
      REAL data1(NPTS),data2(MPTS),gasdev,prob,shift,t
C     generate Gaussian distributed data
      idum=-5
      do 11 i=1,NPTS
```

```
         data1(i)=gasdev(idum)
11       continue
         do 12 i=1,MPTS
           data2(i)=(NSHFT/2.0)*EPS+gasdev(idum)
12       continue
         write(*,'(/1x,t4,a,t18,a,t25,a)') 'Shift','T','Probability'
         do 14 i=1,NSHFT+1
           call ttest(data1,NPTS,data2,MPTS,t,prob)
           shift=(i-1)*EPS
           write(*,'(1x,f6.2,2f12.2)') shift,t,prob
           do 13 j=1,NPTS
             data1(j)=data1(j)+EPS
13         continue
14       continue
         END
```

avevar is an auxiliary routine for ttest. It finds the average and variance of a data set. Sample program xavevar generates a series of Gaussian distributions for $i=1,..,11$, and gives each a shift of $(i-1)$EPS and a variance of i^2. This progression allows you easily to check the operation of avevar "by eye".

```
         PROGRAM xavevar
C        driver for routine avevar
         INTEGER NPTS
         REAL EPS
         PARAMETER(NPTS=1000, EPS=0.1)
         INTEGER i,idum,j
         REAL ave,gasdev,shift,var,data(NPTS)
C        generate Gaussian distributed data
         idum=-5
         write(*,'(1x,t4,a,t14,a,t26,a)') 'Shift','Average','Variance'
         do 12 i=1,11
           shift=(i-1)*EPS
           do 11 j=1,NPTS
             data(j)=shift+i*gasdev(idum)
11         continue
           call avevar(data,NPTS,ave,var)
           write(*,'(1x,f6.2,2f12.2)') shift,ave,var
12       continue
         END
```

tutest also does Student's *t*-test, but applies to the comparison of means of two distributions with different variance. The example xtutest employs the comparison method used in ttest but gives the two distributions data1 and data2 variances of 1.0 and 4.0 respectively.

```
         PROGRAM xtutest
C        driver for routine tutest
         INTEGER NPTS,NSHFT,MPTS
         REAL EPS,VAR1,VAR2
         PARAMETER(NPTS=5000,MPTS=1000,EPS=0.02,VAR1=1.0,
     *       VAR2=4.0,NSHFT=10)
         INTEGER i,idum,j
         REAL data1(NPTS),data2(MPTS),fctr1,fctr2,gasdev,prob,shift,t
C        generate two Gaussian distributions of different variance
         idum=-51773
         fctr1=sqrt(VAR1)
```

```
      do 11 i=1,NPTS
        data1(i)=fctr1*gasdev(idum)
11    continue
      fctr2=sqrt(VAR2)
      do 12 i=1,MPTS
        data2(i)=(NSHFT/2.0)*EPS+fctr2*gasdev(idum)
12    continue
      write(*,'(1x,a,f6.2)') 'Distribution #1 : variance = ',VAR1
      write(*,'(1x,a,f6.2/)') 'Distribution #2 : variance = ',VAR2
      write(*,'(1x,t4,a,t18,a,t25,a)') 'Shift','T','Probability'
      do 14 i=1,NSHFT+1
        call tutest(data1,NPTS,data2,MPTS,t,prob)
        shift=(i-1)*EPS
        write(*,'(1x,f6.2,2f12.2)') shift,t,prob
        do 13 j=1,NPTS
          data1(j)=data1(j)+EPS
13      continue
14    continue
      END
```

tptest goes a step further, and compares two distributions not only having different variances, but also perhaps having point by point correlations. The example xtptest creates two situations, one with correlated and one with uncorrelated distributions. It does this by way of three data sets. data1 is a simple Gaussian distribution of zero mean and unit variance. data2 is data1 plus some additional Gaussian fluctuations of smaller amplitude. data3 is similar to data2 but generated with independent calls to gasdev so that its fluctuations ought not to have any correlation with those of data1. data1 is then given an offset with respect to the others and they are successively shifted as in previous routines. At each step of the shift tptest was applied. Our results are given below:

	Correlated:		Uncorrelated:	
Shift	T	Probability	T	Probability
0.01	3.1199	0.0019	0.6287	0.5298
0.02	2.3399	0.0197	0.4716	0.6374
0.03	1.5600	0.1194	0.3144	0.7534
0.04	0.7800	0.4358	0.1572	0.8751
0.05	0.0000	1.0000	0.0000	1.0000
0.06	-0.7800	0.4358	-0.1572	0.8751
0.07	-1.5600	0.1194	-0.3144	0.7534
0.08	-2.3399	0.0197	-0.4716	0.6374
0.09	-3.1199	0.0019	-0.6287	0.5298
0.10	-3.8999	0.0001	-0.7859	0.4323
0.11	-4.6799	0.0000	-0.9431	0.3460

```
      PROGRAM xtptest
C     driver for routine tptest
      INTEGER NPTS,NSHFT
      REAL ANOISE,EPS
      PARAMETER(NPTS=500,EPS=0.01,NSHFT=10,ANOISE=0.3)
      INTEGER i,idum,j
      REAL ave1,ave2,ave3,gasdev,offset,prob1,prob2,shift,
     *    t1,t2,var1,var2,var3
      REAL data1(NPTS),data2(NPTS),data3(NPTS)
      idum=-5
      write(*,'(1x,t18,a,t46,a)') 'Correlated:','Uncorrelated:'
```

```
      write(*,'(1x,t4,a,t18,a,t25,a,t46,a,t53,a)')
     *      'Shift','T','Probability','T','Probability'
      offset=(NSHFT/2)*EPS
      do 11 j=1,NPTS
        data1(j)=gasdev(idum)
        data2(j)=data1(j)+ANOISE*gasdev(idum)
        data3(j)=gasdev(idum)
        data3(j)=data3(j)+ANOISE*gasdev(idum)
11    continue
      call avevar(data1,NPTS,ave1,var1)
      call avevar(data2,NPTS,ave2,var2)
      call avevar(data3,NPTS,ave3,var3)
      do 12 j=1,NPTS
        data1(j)=data1(j)-ave1+offset
        data2(j)=data2(j)-ave2
        data3(j)=data3(j)-ave3
12    continue
      do 14 i=1,NSHFT+1
        shift=i*EPS
        do 13 j=1,NPTS
          data2(j)=data2(j)+EPS
          data3(j)=data3(j)+EPS
13      continue
        call tptest(data1,data2,NPTS,t1,prob1)
        call tptest(data1,data3,NPTS,t2,prob2)
        write(*,'(1x,f6.2,2x,2f12.4,4x,2f12.4)')
     *        shift,t1,prob1,t2,prob2
14    continue
      END
```

The F-test (subroutine ftest) is a test for differing variances between two distributions. For demonstration purposes, sample program xftest generates two distributions data1 and data2 having Gaussian distributions of unit variance. The values of a third array data3 are then set by multiplying data2 by a series of values factr which takes its variance from 1.0 to 1.1 in ten equal steps. The effect of this on the F-test can be evaluated from the probabilities prob.

```
      PROGRAM xftest
C     driver for routine ftest
      INTEGER NPTS,NVAL,MPTS
      REAL EPS
      PARAMETER(NPTS=1000,MPTS=500,EPS=0.01,NVAL=10)
      INTEGER i,idum,j
      REAL f,factor,gasdev,prob,var
      REAL data1(NPTS),data2(MPTS),data3(MPTS)
C     generate two Gaussian distributions with
C     different variances
      idum=-13
      do 11 j=1,NPTS
        data1(j)=gasdev(idum)
11    continue
      do 12 j=1,MPTS
        data2(j)=gasdev(idum)
12    continue
      write(*,'(1x,t5,a,f5.2)') 'Variance 1 = ',1.0
      write(*,'(1x,t5,a,t21,a,t30,a)')
```

```
      *     'Variance 2','Ratio','Probability'
            do 14 i=1,NVAL+1
              var=1.0+(i-1)*EPS
              factor=sqrt(var)
              do 13 j=1,MPTS
                data3(j)=factor*data2(j)
13            continue
              call ftest(data1,NPTS,data3,MPTS,f,prob)
              write(*,'(1x,f11.4,2x,2f12.4)') var,f,prob
14          continue
            END
```

chsone and chstwo compare two distributions on the basis of a χ^2 test to see if they are different. chsone, specifically, compares a data distribution to an expected distribution. Sample program xchsone generates an exponential distribution bins(i) of data using routine expdev. It then creates an array ebins(i) which is the expected result (a smooth exponential decay in the absence of statistical fluctuations). ebins and bins are compared by chsone to give χ^2 and a probability that they represent the same distribution.

```
            PROGRAM xchsone
C           driver for routine chsone
            INTEGER NBINS,NPTS
            PARAMETER(NBINS=10,NPTS=2000)
            INTEGER i,ibin,idum,j
            REAL chsq,df,expdev,prob,x,bins(NBINS),ebins(NBINS)
            idum=-15
            do 11 j=1,NBINS
              bins(j)=0.0
11          continue
            do 12 i=1,NPTS
              x=expdev(idum)
              ibin=x*NBINS/3.0+1
              if (ibin.le.NBINS) bins(ibin)=bins(ibin)+1.0
12          continue
            do 13 i=1,NBINS
              ebins(i)=3.0*NPTS/NBINS*exp(-3.0*(i-0.5)/NBINS)
13          continue
            call chsone(bins,ebins,NBINS,0,df,chsq,prob)
            write(*,'(1x,t10,a,t25,a)') 'Expected','Observed'
            do 14 i=1,NBINS
              write(*,'(1x,2f15.2)') ebins(i),bins(i)
14          continue
            write(*,'(/1x,t9,a,e12.4)') 'Chi-squared:',chsq
            write(*,'(1x,t9,a,e12.4)') 'Probability:',prob
            END
```

chstwo compares two binned distributions bins1 and bins2, again using a χ^2 test. Sample program xchstwo prepares these distributions both in the same way. Each is composed of 2000 random numbers, drawn from an exponential deviate, and placed into 10 bins. The two data sets are then analyzed by chstwo to calculate χ^2 and probability prob.

```
            PROGRAM xchstwo
C           driver for routine chstwo
            INTEGER NBINS,NPTS
```

```
      PARAMETER(NBINS=10,NPTS=2000)
      INTEGER i,ibin,idum,j
      REAL chsq,df,expdev,prob,x,bins1(NBINS),bins2(NBINS)
      idum=-17
      do 11 j=1,NBINS
        bins1(j)=0.0
        bins2(j)=0.0
11    continue
      do 12 i=1,NPTS
        x=expdev(idum)
        ibin=x*NBINS/3.0+1
        if (ibin.le.NBINS) bins1(ibin)=bins1(ibin)+1.0
        x=expdev(idum)
        ibin=x*NBINS/3.0+1
        if (ibin.le.NBINS) bins2(ibin)=bins2(ibin)+1.0
12    continue
      call chstwo(bins1,bins2,NBINS,0,df,chsq,prob)
      write(*,'(1x,t10,a,t25,a)') 'Dataset 1','Dataset 2'
      do 13 i=1,NBINS
        write(*,'(1x,2f15.2)') bins1(i),bins2(i)
13    continue
      write(*,'(/1x,t10,a,e12.4)') 'Chi-squared:',chsq
      write(*,'(1x,t10,a,e12.4)') 'Probability:',prob
      END
```

The Kolmogorov-Smirnov test used in **ksone** and **kstwo** applies to unbinned distributions with a single independent variable. **ksone** uses the K-S criterion to compare a single data set to an expected distribution, and **kstwo** uses it to compare two data sets. Sample program **xksone** creates data sets with Gaussian distributions and with stepwise increasing variance, and compares their cumulative distribution function to the expected result for a Gaussian distribution of unit variance. This result is the error function and is generated by routine **erf**. Increasing variance in the test distribution should reduce the likelihood that it was drawn from the same distribution represented by the comparison function.

```
      PROGRAM xksone
C     driver for routine ksone
      INTEGER NPTS
      REAL EPS
      PARAMETER(NPTS=1000,EPS=0.1)
      INTEGER i,idum,j
      REAL d,data(NPTS),factr,gasdev,prob,var
      EXTERNAL func
      idum=-5
      write(*,'(/1x,t5,a,t24,a,t44,a/)')
     *    'Variance Ratio','K-S Statistic','Probability'
      do 12 i=1,11
        var=1.0+(i-1)*EPS
        factr=sqrt(var)
        do 11 j=1,NPTS
          data(j)=factr*abs(gasdev(idum))
11      continue
        call ksone(data,NPTS,func,d,prob)
        write(*,'(1x,f14.6,f18.6,e20.4)') var,d,prob
12    continue
      END
```

```
REAL FUNCTION func(x)
REAL erf,x,y
y=x/sqrt(2.0)
func=erf(y)
END
```

kstwo compares the cumulative distribution functions of two unbinned data sets, data1 and data2. In sample program xkstwo, they are both Gaussian distributions, but data2 is given a stepwise increase of variance. In other respects, xkstwo is like xksone.

```
      PROGRAM xkstwo
C     driver for routine kstwo
      INTEGER N1,N2
      REAL EPS
      PARAMETER(N1=2000,N2=1000,EPS=0.1)
      INTEGER i,idum,j
      REAL d,data1(N1),data2(N2),factr,gasdev,prob,var
      idum=-1357
      do 11 j=1,N1
        data1(j)=gasdev(idum)
11    continue
      write(*,'(/1x,t6,a,t26,a,t46,a/)')
     *      'Variance Ratio','K-S Statistic','Probability'
      do 13 i=1,11
        var=1.0+(i-1)*EPS
        factr=sqrt(var)
        do 12 j=1,N2
          data2(j)=factr*gasdev(idum)
12      continue
        call kstwo(data1,N1,data2,N2,d,prob)
        write(*,'(1x,f15.6,f19.6,e20.4)') var,d,prob
13    continue
      END
```

probks is an auxiliary routine for ksone and kstwo which calculates the function $Q_{ks}(\lambda)$ used to evaluate the probability that the two distributions being compared are the same. There is no independent means of producing this function, so in sample program xprobks we have chosen simply to graph it. Our output is reproduced below.

```
      PROGRAM xprobks
C     driver for routine probks
      INTEGER i,j,npts
      REAL alam,eps,probks,scale,value
      CHARACTER text(50)*1
      write(*,*) 'Probability func. for Kolmogorov-Smirnov statistic'
      write(*,'(/1x,t3,a,t15,a,t27,a)') 'Lambda:','Value:','Graph:'
      npts=20
      eps=0.1
      scale=40.0
      do 12 i=1,npts
        alam=i*eps
        value=probks(alam)
        text(1)='*'
        do 11 j=1,50
          if (j.le.nint(scale*value)) then
```

```
         text(j)='*'
      else
         text(j)=' '
      endif
11    continue
      write(*,'(1x,f9.6,f12.6,4x,50a1)') alam,value,
     *     (text(j),j=1,50)
12    continue
      END
```

```
Probability func. for Kolmogorov-Smirnov statistic
 Lambda:    Value:     Graph:
 0.100000   1.000000   ****************************************
 0.200000   1.000000   ****************************************
 0.300000   0.999991   ****************************************
 0.400000   0.997192   ****************************************
 0.500000   0.963945   ***************************************
 0.600000   0.864283   ***********************************
 0.700000   0.711235   ****************************
 0.800000   0.544142   *********************
 0.900000   0.392731   ***************
 1.000000   0.270000   ***********
 1.100000   0.177718   *******
 1.200000   0.112250   ****
 1.300000   0.068092   ***
 1.400000   0.039682   **
 1.500000   0.022218   *
 1.600000   0.011952
 1.700000   0.006177
 1.800000   0.003068
 1.900000   0.001464
 2.000000   0.000671
```

Subroutine cntab1 analyzes a two-dimensional contingency table and returns several parameters describing any association between its nominal variables. Sample program xcntab1 supplies a table from a file TABLE.DAT that is listed in the Appendix to this chapter. The table shows the rate of certain accidents, tabulated on a monthly basis. These data are listed, as well as their statistical properties, by xcntab1. We found the results to be:

Chi-squared	5026.30
Degrees of Freedom	88.00
Probability	.0000
Cramer-V	.0772
Contingency Coeff.	.2134

```
      PROGRAM xcntab1
C     driver for routine cntab1
      INTEGER NDAT,NMON
      PARAMETER(NDAT=9,NMON=12)
      INTEGER i,j,nmbr(NDAT,NMON)
      REAL ccc,chisq,cramrv,df,prob
      CHARACTER fate(NDAT)*15,mon(NMON)*5,text*64
      open(7,file='TABLE.DAT',status='OLD')
      read(7,*)
      read(7,'(a)') text
```

```
      read(7,'(15x,12a5/)') (mon(i),i=1,12)
      do 11 i=1,NDAT
        read(7,'(a15,12i5)') fate(i),(nmbr(i,j),j=1,12)
11    continue
      close(7)
      write(*,'(/1x,a/)') text
      write(*,'(1x,15x,12a5)') (mon(i),i=1,12)
      do 12 i=1,NDAT
        write(*,'(1x,a,12i5)') fate(i),(nmbr(i,j),j=1,12)
12    continue
      call cntab1(nmbr,NDAT,NMON,chisq,df,prob,cramrv,ccc)
      write(*,'(/1x,a,t20,f20.2)') 'Chi-squared',chisq
      write(*,'(1x,a,t20,f20.2)') 'Degrees of Freedom',df
      write(*,'(1x,a,t20,f20.4)') 'Probability',prob
      write(*,'(1x,a,t20,f20.4)') 'Cramer-V',cramrv
      write(*,'(1x,a,t20,f20.4)') 'Contingency Coeff.',ccc
      END
```

The test looks for any association between accidents and the months in which they occur. TABLE.DAT clearly shows some. Drownings, for example, happen mostly in the summer. cntab2 carries out a similar analysis on TABLE.DAT but measures associations on the basis of entropy. Sample program xcntab2 prints out the following entropies for the table:

```
      Entropy of Table              4.0368
      Entropy of x-distribution     1.5781
      Entropy of y-distribution     2.4820
      Entropy of y given x          2.4588
      Entropy of x given y          1.5548
      Dependency of y on x           .0094
      Dependency of x on y           .0147
      Symmetrical dependency         .0114
```

```
      PROGRAM xcntab2
C     driver for routine cntab2
      INTEGER NI,NMON
      PARAMETER(NI=9,NMON=12)
      INTEGER i,j,nmbr(NI,NMON)
      REAL h,hx,hxgy,hy,hygx,uxgy,uxy,uygx
      CHARACTER fate(NI)*15,mon(NMON)*5,text*64
      open(7,file='TABLE.DAT',status='OLD')
      read(7,*)
      read(7,'(a)') text
      read(7,'(15x,12a5/)') (mon(i),i=1,12)
      do 11 i=1,NI
        read(7,'(a15,12i5)') fate(i),(nmbr(i,j),j=1,12)
11    continue
      close(7)
      write(*,'(/1x,a/)') text
      write(*,'(1x,15x,12a5)') (mon(i),i=1,12)
      do 12 i=1,NI
        write(*,'(1x,a,12i5)') fate(i),(nmbr(i,j),j=1,12)
12    continue
      call cntab2(nmbr,NI,NMON,h,hx,hy,hygx,hxgy,uygx,uxgy,uxy)
      write(*,'(/1x,a,t30,f10.4)') 'Entropy of Table',h
      write(*,'(1x,a,t30,f10.4)') 'Entropy of x-distribution',hx
      write(*,'(1x,a,t30,f10.4)') 'Entropy of y-distribution',hy
```

```
      write(*,'(1x,a,t30,f10.4)')') 'Entropy of y given x',hygx
      write(*,'(1x,a,t30,f10.4)')') 'Entropy of x given y',hxgy
      write(*,'(1x,a,t30,f10.4)')') 'Dependency of y on x',uygx
      write(*,'(1x,a,t30,f10.4)')') 'Dependency of x on y',uxgy
      write(*,'(1x,a,t30,f10.4/)') 'Symmetrical dependency',uxy
      END
```

The dependencies of x on y and y on x indicate the degree to which the type of accident can be predicted by knowing the month, or vice-versa.

pearsn makes an examination of two ordinal or continuous variables to find linear correlations. It returns a linear correlation coefficient r, a probability of correlation prob, and Fisher's z. Sample program xpearsn sets up data pairs in arrays dose and spore which show hypothetical data for the spore count from plants exposed to various levels of γ-rays. The results of applying pearsn to this data set are compared with the correct results by the program.

```
      PROGRAM xpearsn
C     driver for routine pearsn
      INTEGER i
      REAL prob,r,z
      REAL dose(10),spore(10)
      DATA dose/56.1,64.1,70.0,66.6,82.,91.3,90.,99.7,115.3,110./
      DATA spore/0.11,0.4,0.37,0.48,0.75,0.66,0.71,1.2,1.01,0.95/
      write(*,'(1x,a)')
     *    'Effect of Gamma Rays on Man-in-the-Moon Marigolds'
      write(*,'(1x,a,t29,a)') 'Count Rate (cpm)','Pollen Index'
      do 11 i=1,10
        write(*,'(1x,f10.2,f25.2)') dose(i),spore(i)
11    continue
      call pearsn(dose,spore,10,r,prob,z)
      write(*,'(/1x,t24,a,t38,a)') 'PEARSN','Expected'
      write(*,'(1x,a,t18,2e15.6)') 'Corr. Coeff.',r,0.906959
      write(*,'(1x,a,t18,2e15.6)') 'Probability',prob,0.292650e-3
      write(*,'(1x,a,t18,2e15.6/)') 'Fisher''s Z',z,1.51011
      END
```

Rank order correlation may be done with spear to compare two distributions data1 and data2 for correlation. Correlations are reported both in terms of d, the sum-squared difference in ranks, and rs, Spearman's rank correlation parameter. Sample program xspear applies the calculation to the data in table TABLE2.DAT (see Appendix) which shows the solar flux incident on various cities during different months of the year. It then checks for correlations between columns of the table, considering each column as a separate data set. In this fashion it looks for correlations between the July solar flux and that of other months. The probability of such correlations are shown by probd and probrs. Our results are:

```
Check correlation of sampled U.S. solar radiation (july with other months)
Month      D       St. Dev.     PROBD       Spearman R    PROBRS
 jul      0.00    -4.358899    0.000013     1.000000     0.000000
 aug    122.00    -3.958458    0.000075     0.908132     0.000000
 sep    218.00    -3.643896    0.000269     0.835967     0.000004
 oct    384.00    -3.098495    0.001945     0.710843     0.000443
 nov    390.50    -3.077642    0.002086     0.706060     0.000503
 dec    622.00    -2.318075    0.020445     0.531803     0.015806
 jan    644.50    -2.244251    0.024816     0.514866     0.020181
```

feb	483.50	-2.772503	0.005563	0.636056	0.002573
mar	497.00	-2.728208	0.006368	0.625894	0.003158
apr	405.50	-3.027925	0.002462	0.694654	0.000677
may	264.00	-3.492371	0.000479	0.801205	0.000022
jun	121.50	-3.960099	0.000075	0.908509	0.000000

```
      PROGRAM xspear
C     driver for routine spear
      INTEGER NDAT,NMON
      PARAMETER(NDAT=20,NMON=12)
      INTEGER i,j
      REAL d,probd,probrs,rs,zd
      REAL data1(NDAT),data2(NDAT),rays(NDAT,NMON)
      REAL wksp1(NDAT),wksp2(NDAT),ave(NDAT),zlat(NDAT)
      CHARACTER city(NDAT)*15,mon(NMON)*4,text*64
      open(7,file='TABLE2.DAT',status='OLD')
      read(7,*)
      read(7,'(a)') text
      read(7,'(15x,12a4/)') (mon(i),i=1,12)
      do 11 i=1,NDAT
        read(7,'(a15,12f4.0,f6.0,f6.1)')
     *      city(i),(rays(i,j),j=1,12),ave(i),zlat(i)
11    continue
      close(7)
      write(*,*) text
      write(*,'(1x,15x,12a4)') (mon(i),i=1,12)
      do 12 i=1,NDAT
        write(*,'(1x,a,12i4,i6,f6.1)') city(i),
     *      (nint(rays(i,j)),j=1,12)
12    continue
C     check temperature correlations between different months
      write(*,'(/1x,a)')
     *      'Are sunny summer places also sunny winter places?'
      write(*,'(1x,2a)') 'Check correlation of sampled U.S. solar ',
     *      'radiation (july with other months)'
      write(*,'(/1x,a,t16,a,t23,a,t37,a,t49,a,t63,a/)')
     *      'Month','D','St. Dev.','PROBD',
     *      'Spearman R','PROBRS'
      do 13 i=1,NDAT
        data1(i)=rays(i,1)
13    continue
      do 15 j=1,12
        do 14 i=1,NDAT
          data2(i)=rays(i,j)
14      continue
        call spear(data1,data2,NDAT,wksp1,wksp2,d,zd,probd,rs,probrs)
        write(*,'(1x,a,f13.2,2f12.6,3x,2f12.6)')
     *      mon(j),d,zd,probd,rs,probrs
15    continue
      END
```

crank is an auxiliary routine for spear and is used in conjunction with sort2. The latter sorts an array, and crank then assigns ranks to each data entry, including the midranking of ties. Sample program xcrank uses the solar flux data of TABLE2.DAT (see Appendix) to illustrate. Each column of the solar flux table is replaced by the rank order of its entries. You can check the rank order chart against the chart of original values to verify the ordering.

```
      PROGRAM xcrank
C     driver for routine crank
      INTEGER NDAT,NMON
      PARAMETER(NDAT=20,NMON=12)
      INTEGER i,j
      REAL s,data(NDAT),rays(NDAT,NMON),order(NDAT),ave(NDAT),zlat(NDAT)
      CHARACTER city(NDAT)*15,mon(NMON)*4,text*64
      open(7,file='TABLE2.DAT',status='OLD')
      read(7,*)
      read(7,'(a)') text
      read(7,'(15x,12a4/)') (mon(i),i=1,NMON)
      do 11 i=1,NDAT
        read(7,'(a15,12f4.0,f6.0,f6.1)')
     *       city(i),(rays(i,j),j=1,NMON),ave(i),zlat(i)
11    continue
      close(7)
      write(*,'(1x,a)') text
      write(*,'(1x,15x,12a4)') (mon(i),i=1,NMON)
      do 12 i=1,NDAT
        write(*,'(1x,a,12i4)')
     *       city(i),(nint(rays(i,j)),j=1,NMON)
12    continue
      write(*,'(/1x,a)') 'press RETURN to continue...'
      read(*,*)
C     Replace solar flux in each column by rank order
      do 15 j=1,NMON
        do 13 i=1,NDAT
          data(i)=rays(i,j)
          order(i)=i
13      continue
        call sort2(NDAT,data,order)
        call crank(NDAT,data,s)
        do 14 i=1,NDAT
          rays(nint(order(i)),j)=data(i)
14      continue
15    continue
      write(*,'(1x,t17,12a4)') (mon(i),i=1,NMON)
      do 16 i=1,NDAT
      write(*,'(1x,a,12i4)') city(i),(nint(rays(i,j)),j=1,NMON)
16    continue
      END
```

kendl1 and kendl2 test for monotonic correlations of ordinal data. They differ in that kendl1 compares two data sets of the same rank, while kendl2 operates on a contingency table. Sample program xkendl1, for example, uses kendl1 to look for pair correlations in our five random number routines. That is to say, it tests for randomness by seeing if two consecutive numbers from the generator have a monotonic correlation. It uses the random number generators ran0, ..., ran4, one at a time, to generate 200 pairs of random numbers each. Then kendl1 tests for correlation of the pairs, and a chart is made showing Kendall's τ, the standard deviation from the null hypotheses, and the probability. For a better test of the generators, you may wish to increase the number of pairs NDAT. It would also be a good idea to see how your result depends on the value of the seed idum.

```
         PROGRAM xkendl1
C        driver for routine kendl1
         INTEGER NDAT
         PARAMETER(NDAT=200)
         INTEGER i,idum,j
         REAL prob,ran0,ran1,ran2,ran3,ran4,tau,z
         REAL data1(NDAT),data2(NDAT)
         CHARACTER text(5)*4
         DATA text/'RAN0','RAN1','RAN2','RAN3','RAN4'/
         write(*,'(/1x,a/)') 'Pair correlations of RAN0 ... RAN4'
         write(*,'(2x,a,t16,a,t34,a,t50,a,/)')
     *      'Program','Kendall Tau','Std. Dev.','Probability'
         do 12 i=1,5
            idum=-1357
            do 11 j=1,NDAT
               if (i.eq.1) then
                  data1(j)=ran0(idum)
                  data2(j)=ran0(idum)
               else if (i.eq.2) then
                  data1(j)=ran1(idum)
                  data2(j)=ran1(idum)
               else if (i.eq.3) then
                  data1(j)=ran2(idum)
                  data2(j)=ran2(idum)
               else if (i.eq.4) then
                  data1(j)=ran3(idum)
                  data2(j)=ran3(idum)
               else if (i.eq.5) then
                  data1(j)=ran4(idum)
                  data2(j)=ran4(idum)
               endif
11          continue
            call kendl1(data1,data2,NDAT,tau,z,prob)
            write(*,'(1x,t4,a,3f17.6)') text(i),tau,z,prob
12       continue
         END
```

Here is the output from the above test:

```
Pair correlations of RAN0 ... RAN4
  Program       Kendall Tau          Std. Dev.        Probability
  RAN0          -0.010452           -0.219801          0.826026
  RAN1          -0.023719           -0.498780          0.617934
  RAN2           0.130251            2.739064          0.006161
  RAN3          -0.062814           -1.320922          0.186527
  RAN4           0.023015            0.483986          0.628396
```

Sample program xkendl2, for subroutine kendl2, prepares a contingency table based on the routines irbit1 and irbit2. You may recall that these routines generate random binary sequences. The program checks the sequences by breaking them into groups of three bits. Each group is treated as a three-bit binary number. Two consecutive groups then act as indices into an 8×8 contingency table that records how many times each possible sequence of six bits (two groups) occurs. For each random bit generator, NDAT=1000 samples are taken. Then the contingency table tab(k,l) is analyzed by kendl2 to find Kendall's τ, the standard deviation, and the probability. Notice that Kendall's τ can only be applied when both variables are ordinal (here, the numbers 0 to 7), and that the test is specifically for monotonic correlations. In this case we are actually testing whether the larger 3-bit binary numbers tend to be followed by others of their own kind. Within the program, we have expressed this roughly as a test of whether ones or zeros tend to come in groups more than they should.

```
      PROGRAM xkendl2
C     driver for routine kendl2
      INTEGER IP,JP,NDAT
      PARAMETER(NDAT=1000,IP=8,JP=8)
      INTEGER i,ifunc,irbit1,irbit2,iseed,j,k,l,m,n
      REAL prob,tau,z,tab(IP,JP)
      CHARACTER text(8)*3
      DATA text/'000','001','010','011','100','101','110','111'/
      write(*,*) 'Are ones followed by zeros and vice-versa?'
      i=IP
      j=JP
      do 17 ifunc=1,2
        iseed=2468
        write(*,'(/1x,a,i1/)') 'Test of IRBIT',ifunc
        do 12 k=1,i
          do 11 l=1,j
            tab(k,l)=0.0
11        continue
12      continue
        do 15 m=1,NDAT
          k=1
          do 13 n=0,2
            if (ifunc.eq.1) then
              k=k+irbit1(iseed)*(2**n)
            else
              k=k+irbit2(iseed)*(2**n)
            endif
13        continue
          l=1
          do 14 n=0,2
            if (ifunc.eq.1) then
              l=l+irbit1(iseed)*(2**n)
            else
              l=l+irbit2(iseed)*(2**n)
            endif
14        continue
          tab(k,l)=tab(k,l)+1.0
15      continue
        call kendl2(tab,i,j,IP,JP,tau,z,prob)
        write(*,'(4x,8a6/)') (text(n),n=1,8)
        do 16 n=1,8
          write(*,'(1x,a,8i6)') text(n),(nint(tab(n,m)),m=1,8)
16      continue
        write(*,'(/7x,a,t24,a,t38,a)') 'Kendall Tau','Std. Dev.',
     *      'Probability'
        write(*,'(1x,3f15.6/)') tau,z,prob
        write(*,*) 'Press RETURN to continue ...'
        read(*,*)
17    continue
      END
```

The output of `xkendl2` looks like this:

```
Are ones followed by zeros and vice-versa?
Test of IRBIT1
        000   001   010   011   100   101   110   111
000      17    15     3    14    14    18    10    15
001      18    14     8    16    14    22    17    10
010      18    15     6    14    14    16    13    15
011      17    13     8    18    21    26    17    13
100      15    16    12    11    18    18    18    18
101      14    10    15    17    18    16    25    15
110      16    12    25    15    23    15    13    22
111      17    15    16    18    16    15    23    14
         Kendall Tau      Std. Dev.      Probability
          0.011227        0.531589        0.595010
Test of IRBIT2
        000   001   010   011   100   101   110   111
000      21    12    21    16    17     9    14    15
001      14    17    15    18    10    17    20    16
010      19    17    15    16    17    17     9    14
011      12    18     9    16    13    26    13    13
100      16    18    14    16    12    19    15    11
101      12    21    20    16    17    13     8    15
110      14    18    20    17    12    17    12    19
111      19    16    15    22    19     9    19    13
         Kendall Tau      Std. Dev.      Probability
         -0.003014       -0.142715        0.886516
```

The routine `ks2d1s` tests a single sample (hence "1s") of a two-dimensional distribution (hence "2d") against a theoretical model distribution, returning the significance level with which the distribution can be said to differ from the model. The model, supplied by the user, is contained in the routine `quadvl`. The main book supplies a sample `quadvl` where the model is a distribution that is uniform inside the square $|x| < 1$, $|y| < 1$, so we will use that `quadvl` in our example program.

We generate a sample that is made nonuniform by "warping" the output of the uniform random number generator by a quadratic function of adjustable amplitude (`factor`). Then, the program invokes `ks2d1s` and prints the significance of the result. This is repeated some number of times, so that you can get an idea of how variable the test is from run to run.

If you run the program, you will see that `factor` as large as 0.1 is easily detectable in a 500 point sample (i.e., it gives very small values for `PROB`), but not detectable in 100 points. In a 100 point sample, `factor` must be on the order of 0.5 before it becomes easily detectable.

```
      PROGRAM xks2d1s
C     driver for routine ks2d1s
      INTEGER NMAX
      PARAMETER(NMAX=1000)
      INTEGER idum,j,jtrial,n1,ntrial
      REAL d,ran1,prob,factor,u,v,x1(NMAX),y1(NMAX)
      EXTERNAL quadvl
1     write(*,*) 'HOW MANY POINTS?'
      read(*,*,END=999) n1
      if (n1.gt.NMAX) then
```

```
        write(*,*) 'n1 too large.'
        goto 1
      endif
2     write(*,*) 'WHAT FACTOR NONLINEARITY (0 to 1)?'
      read(*,*,END=999) factor
      if (factor.lt.0.) then
        write(*,*) 'factor less than 0'
        goto 2
      endif
      if (factor.gt.1.) then
        write(*,*) 'factor greater than 1'
        goto 2
      endif
      write(*,*) 'HOW MANY TRIALS?'
      read(*,*,END=999) ntrial
      idum=-289-ntrial-n1
      do 12 jtrial=1,ntrial
        do 11 j=1,n1
          u=ran1(idum)
          u=u*((1.-factor)+u*factor)
          x1(j)=2.*u-1.
          v=ran1(idum)
          v=v*((1.-factor)+v*factor)
          y1(j)=2.*v-1.
11      continue
        call ks2d1s(x1,y1,n1,quadvl,d,prob)
        write(*,'(1x,a7,2f12.6)') 'D,PROB= ',d,prob
12    continue
      goto 1
999   write(*,*) 'NORMAL COMPLETION'
      STOP
      END
```

The routine ks2d2s performs a similar statistical test, but on two samples and without the necessity of a model. The example program is quite similar to that for ks2d1s. Now, however, we generate two samples, drawn from 2-dimensional Gaussian distributions of slightly different shape (as controlled by the variable shrink).

Running the program, one finds that with, e.g., sample sizes of 100 points for the first sample and 80 points for the second sample, a value shrink $= 0.5$ is easily detectable as a low value for prob, while shrink $= 0.8$ is not.

```
      PROGRAM xks2d2s
C     driver for routine ks2d2s
      INTEGER NMAX
      PARAMETER(NMAX=1000)
      INTEGER idum,j,jtrial,n1,n2,ntrial
      REAL d,gasdev,prob,shrink,u,v,x1(NMAX),y1(NMAX),x2(NMAX),y2(NMAX)
1     write(*,*) 'INPUT N1,N2'
      read(*,*,END=999) n1,n2
      if (n1.gt.NMAX) then
        write(*,*) 'n1 too large.'
        goto 1
      endif
      if (n2.gt.NMAX) then
        write(*,*) 'n2 too large.'
        goto 1
```

```
          endif
          write(*,*) 'WHAT SHRINKAGE?'
          read(*,*,END=999) shrink
          write(*,*) 'HOW MANY TRIALS?'
          read(*,*,END=999) ntrial
          if (ntrial.gt.NMAX) then
            write(*,*) 'Too many trials.'
            goto 1
          endif
          idum=-287-ntrial-n1-n2
          do 13 jtrial=1,ntrial
            do 11 j=1,n1
              u=gasdev(idum)
              v=gasdev(idum)*shrink
              x1(j)=u+v
              y1(j)=u-v
11          continue
            do 12 j=1,n2
              u=gasdev(idum)*shrink
              v=gasdev(idum)
              x2(j)=u+v
              y2(j)=u-v
12          continue
            call ks2d2s(x1,y1,n1,x2,y2,n2,d,prob)
            write(*,'(1x,a7,2f12.6)') 'D,PROB= ',d,prob
13        continue
          goto 1
999       write(*,*) 'NORMAL COMPLETION'
          STOP
          END
```

Sample program `xsavgol` tests `savgol` by computing Savitzky-Golay smoothing coefficients for various values of M, n_L and n_R. It then compares the answers with pre-saved values.

```
          PROGRAM xsavgol
C         driver for routine savgol
          INTEGER NMAX,NTEST
          PARAMETER(NMAX=1000,NTEST=6)
          INTEGER i,j,m,nl,np,nr,nltest(NTEST),nrtest(NTEST),mtest(NTEST)
          REAL c(NMAX),sum
          CHARACTER*39 ans(2*NTEST)
          DATA mtest /2,2,2,2,4,4/
          DATA nltest /2,3,4,5,4,5/
          DATA nrtest /2,1,0,5,4,5/
          DATA ans /'                     -0.086  0.343 ',
     *    ' 0.486  0.343 -0.086',
     *    '              -0.143  0.171  0.343 ','0.371  0.257',
     *    '         0.086 -0.143 -0.086  0.257 0','.886',
     *    ' -0.084  0.021  0.103  0.161  0.196  0.',
     *    '207  0.196  0.161  0.103  0.021 -0.084',
     *    '              0.035 -0.128  0.070  0.315 0',
     *    '.417  0.315  0.070 -0.128  0.035',
     *    '  0.042 -0.105 -0.023  0.140  0.280  0.',
     *    '333  0.280  0.140 -0.023 -0.105  0.042'/
          write(*,*) 'M nl nr'
          write(*,'(t24,a)') 'Sample Savitzky-Golay Coefficients'
```

```
      do 13 i=1,NTEST
        m=mtest(i)
        nl=nltest(i)
        nr=nrtest(i)
        np=nl+nr+1
        call savgol(c,np,nl,nr,0,m)
        sum=0.
        do 11 j=1,np
          sum=sum+c(j)
11      continue
        write(*,'(1x,3i2)') m,nl,nr
        write(*,'(1x,a2,$)') ' '
        do 12 j=nl,4
          write(*,'(1x,a7,$)') ' '
12      continue
        write(*,'(11f7.3)') (c(j),j=nl+1,1,-1),(c(np-j+1),j=1,nr)
        write(*,'(1x,a6,f7.3)') 'Sum = ',sum
        write(*,'(1x,a12,/,4x,t4,a39,t43,a39)') 'Compare ans:',
     *        ans(2*i-1),ans(2*i)
13    continue
      END
```

Appendix

File TABLE.DAT:

Accidental Deaths by Month and Type (1979)

Month:	jan	feb	mar	apr	may	jun	jul	aug	sep	oct	nov	dec
Motor Vehicle	3298	3304	4241	4291	4594	4710	4914	4942	4861	4914	4563	4892
Falls	1150	1034	1089	1126	1142	1100	1112	1099	1114	1079	999	1181
Drowning	180	190	370	530	800	1130	1320	990	580	320	250	212
Fires	874	768	630	516	385	324	277	272	271	381	533	760
Choking	299	264	258	247	273	269	251	269	271	279	297	266
Fire-arms	168	142	122	140	153	142	147	160	162	172	266	230
Poisons	298	277	346	263	253	239	268	228	240	260	252	241
Gas-poison	267	193	144	127	70	63	55	53	60	118	150	172
Other	1264	1234	1172	1220	1547	1339	1419	1453	1359	1308	1264	1246

File TABLE2.DAT:

Average solar radiation (watts/square meter) for selected cities

Month:	jul	aug	sep	oct	nov	dec	jan	feb	mar	apr	may	jun	ave	lat
Atlanta, GA	257	246	201	166	30	102	106	140	184	236	258	271	192	34.0
Barrow, AK	208	123	56	20	0	0	0	18	87	184	248	256	100	71.0
Bismark, ND	296	251	185	132	78	60	76	121	170	217	267	284	178	47.0
Boise, ID	324	275	221	152	88	60	69	113	164	235	284	309	191	43.5
Boston, MA	240	206	165	115	70	58	67	96	142	176	228	242	150	42.5
Caribou, ME	246	218	161	102	53	51	66	111	178	194	229	232	153	47.0
Cleveland, OH	267	239	182	127	68	56	60	87	151	182	253	271	162	41.5
Dodge City, KS	311	287	239	184	138	113	123	153	202	256	275	315	216	38.0
El Paso, TX	324	309	278	224	178	151	160	209	266	317	346	353	260	32.0
Fresno, CA	323	293	243	182	117	77	90	143	212	264	308	337	216	37.0
Greensboro, NC	263	235	197	156	118	95	97	134	171	227	257	273	185	36.0
Honolulu, HI	305	293	271	245	208	176	175	200	234	262	300	297	247	21.0
Little Rock, AR	270	250	214	167	118	91	96	127	173	220	256	272	188	35.0
Miami, FL	260	246	216	188	171	154	166	201	238	263	267	257	219	26.0
New York, NY	251	238	175	127	77	62	71	102	151	183	220	255	159	41.0

Omaha, NE	275	252	192	142	96	80	99	134	172	224	248	272	182	21.0
Rapid City, SD	288	262	208	152	99	76	90	135	193	235	259	287	190	44.0
Seattle, WA	242	209	150	84	44	29	34	60	118	174	216	228	132	47.5
Tucson, AZ	304	286	281	216	172	144	151	195	264	322	358	343	253	41.0
Washington, DC	267	190	196	145	75	64	101	124	153	182	215	247	163	39.0

Chapter 15: Modeling of Data

Chapter 15 of Numerical Recipes deals with the fitting of a model function to a set of data, in order to summarize the data in terms of a few model parameters. Both traditional least-squares fitting and robust fitting are considered. Fits to a straight line are carried out by routine `fit`. *When there are errors in both x and y, straight-line fitting is done by* `fitexy`. *More general linear least-squares fits are handled by* `lfit` *and* `covsrt`. *(Remember that the term "linear" here refers not to a linear dependence of the fitting function on its argument, but rather to the fact that a linear combination of the functions is fit to the data.) In cases where* `lfit` *fails, owing probably to near degeneracy of some basis functions, the answer may still be found using* `svdfit` *and* `svdvar`. *In fact, these are generally recommended in preference to* `lfit` *because they never (?) fail. For nonlinear least-squares fits, the Levenberg-Marquardt method is discussed, and is implemented in* `mrqmin`, *which makes use also of* `covsrt` *and* `mrqcof`.

Robust estimation is discussed in several forms, and illustrated by routine `medfit` *which fits a straight line to data points based on the criterion of least absolute deviations rather than least-squared deviations.* `rofunc` *is an auxiliary function for* `medfit`.

$$\star \quad \star \quad \star \quad \star$$

Routine `fit` fits a set of N data points $(x(i),y(i))$, with standard deviations $sig(i)$, to the linear model $y = Ax + B$. It uses χ^2 as the criterion for goodness-of-fit. To demonstrate `fit`, we generate some noisy data in sample program `xfit`. For `NPT` values of i we take $x = 0.1i$ and $y = -2x + 1$ plus some values drawn from a Gaussian distribution to represent noise. Then we make two calls to `fit`, first performing the fit without allowance for standard deviations $sig(i)$, and then with such allowance. Since $sig(i)$ has been set to the constant value `spread`, it should not affect the resulting parameter values. The values output from this routine are:

Ignoring standard deviation:

```
A =  1.079574     Uncertainty:  0.099821
B = -2.006663     Uncertainty:  0.017161
Chi-squared:      24.047960
Goodness-of-fit:  1.000000
```

Including standard deviation:

```
A =  1.079574     Uncertainty:  0.100755
B = -2.006663     Uncertainty:  0.017321
Chi-squared:      96.191841
```

```
Goodness-of-fit:   0.532771
```

```
      PROGRAM xfit
C     driver for routine fit
      INTEGER NPT
      REAL SPREAD
      PARAMETER(NPT=100,SPREAD=0.5)
      INTEGER i,idum,mwt
      REAL a,b,chi2,gasdev,q,siga,sigb,sig(NPT),x(NPT),y(NPT)
      idum=-117
      do 11 i=1,NPT
        x(i)=0.1*i
        y(i)=-2.0*x(i)+1.0+SPREAD*gasdev(idum)
        sig(i)=SPREAD
11    continue
      do 12 mwt=0,1
        call fit(x,y,NPT,sig,mwt,a,b,siga,sigb,chi2,q)
        if (mwt.eq.0) then
          write(*,'(//1x,a)') 'Ignoring standard deviation'
        else
          write(*,'(//1x,a)') 'Including standard deviation'
        endif
        write(*,'(1x,t5,a,f9.6,t24,a,f9.6)') 'A = ',a,'Uncertainty: ',
     *       siga
        write(*,'(1x,t5,a,f9.6,t24,a,f9.6)') 'B = ',b,'Uncertainty: ',
     *       sigb
        write(*,'(1x,t5,a,4x,f10.6)') 'Chi-squared: ',chi2
        write(*,'(1x,t5,a,f10.6)') 'Goodness-of-fit: ',q
12    continue
      END
```

Here is a program that puts the routine fitexy, which fits a straight line to data with error bars in both x and y, through its paces. The underlying straight line is taken to be

$$y = 2x - 5$$

We generate a synthetic data set that has errors in both x and y, with the errors having a magnitude that is highly variable (but known) from point to point. We then call fitexy and print its result. As a comparison, we set the x errors to zero and do the fit both with fitexy and with the simpler routine fit. Similarly, interchanging the roles of x and y, we can do this setting the y errors to zero. (We have to transform the answers, however, so as to be able to interpret x and y in their true roles.)

Output from this program looks like this:

```
Values of a,b,siga,sigb,chi2,q:
Fit with x and y errors gives:
 -5.058980    2.022800    0.274571    0.020892 42.586609    0.038141
Setting x errors to zero gives:
 -5.538356    2.008888    0.110675    0.009360 277.534546    0.000000
...to be compared with fit result:
 -5.538357    2.008888    0.104930    0.008537 277.534424    0.000000
Setting y errors to zero gives:
 -5.477018    2.043204    0.184778    0.012426 105.700729    0.000000
...to be compared with fit result:
 -5.494896    2.044625    0.160466    0.012380 105.687660    0.000000
```

Notice that the intercept and slope are obtained correctly, notwithstanding a somewhat unlucky value for the χ^2 probability. (This demonstrates that you should not reject a hypothesis just because its χ^2 probability is as small as a few percent: that happens a few percent of the time for correct models!)

Setting either the x or y errors to zero, however, produces significantly wrong results for the intercept, as well as misleading formal errors on the parameters and unacceptable χ^2 probabilities.

```
      PROGRAM xfitexy
C     driver for routine fitexy
      INTEGER NPT
      PARAMETER (NPT=30)
      REAL a,b,chi2,gasdev,q,ran1,sa,sb,siga,sigb
      REAL x(NPT),y(NPT),dx(NPT),dy(NPT),dz(NPT)
      INTEGER idum,j
      DATA dz /NPT*0./
      idum=-1
      do 11 j=1,NPT
        dx(j)=0.1+ran1(idum)
        dy(j)=0.1+ran1(idum)
        x(j)=10.+10.*gasdev(idum)
        y(j)=2.*x(j)-5.+dy(j)*gasdev(idum)
        x(j)=x(j)+dx(j)*gasdev(idum)
11    continue
      write(*,*) 'Values of a,b,siga,sigb,chi2,q:'
      write(*,*) 'Fit with x and y errors gives:'
      call fitexy(x,y,NPT,dx,dy,a,b,siga,sigb,chi2,q)
      write(*,'(6f11.6)') a,b,siga,sigb,chi2,q
      write(*,*)
      write(*,*) 'Setting x errors to zero gives:'
      call fitexy(x,y,NPT,dz,dy,a,b,siga,sigb,chi2,q)
      write(*,'(6f11.6)') a,b,siga,sigb,chi2,q
      write(*,*) '...to be compared with fit result:'
      call fit(x,y,NPT,dy,1,a,b,siga,sigb,chi2,q)
      write(*,'(6f11.6)') a,b,siga,sigb,chi2,q
      write(*,*)
      write(*,*) 'Setting y errors to zero gives:'
      call fitexy(x,y,NPT,dx,dz,a,b,siga,sigb,chi2,q)
      write(*,'(6f11.6)') a,b,siga,sigb,chi2,q
      write(*,*) '...to be compared with fit result:'
      call fit(y,x,NPT,dx,1,a,b,siga,sigb,chi2,q)
      sa=sqrt(siga**2+sigb**2*(a/b)**2)/b
      sb=sigb/b**2
      write(*,'(6f11.6)') -a/b,1./b,sa,sb,chi2,q
      END
```

`lfit` carries out the same sort of fit as `fit` but this time does a linear least-squares fit to a more general function. In sample program `xlfit` the chosen function is a linear sum of powers of x, generated by subroutine `funcs`. For convenience in checking the result we have generated data according to $y = 1 + 2x + 3x^2 + \cdots$. This series is truncated depending on the choice of NTERM, and some Gaussian noise is added to simulate realistic data. The `sig(i)` are taken as constant errors. `lfit` is called twice to fit the same data. The first time `ia(i)` is set to 1, so that `lfit` fits for all the parameters: `a(1)` ≈ 1.0, `a(2)` ≈ 2.0, `a(3)` ≈ 3.0,.... Then, as a test of the ia feature, which determines which parameters are to be fit, the fit is restricted to odd-numbered parameters, while even-numbered parameters are fixed. In this case the elements of the covariance matrix associated with fixed parameters should be zero. In xlfit, we have set NTERM=5 to fit a 5th degree polynomial.

```
      PROGRAM xlfit
C     driver for routine lfit
      INTEGER NPT,NTERM
      REAL SPREAD
      PARAMETER(NPT=100,SPREAD=0.1,NTERM=5)
      INTEGER i,idum,j,ia(NTERM)
      REAL chisq,gasdev
      REAL x(NPT),y(NPT),sig(NPT),a(NTERM),covar(NTERM,NTERM)
      EXTERNAL funcs
      idum=-911
      do 12 i=1,NPT
        x(i)=0.1*i
        call funcs(x(i),a,NTERM)
        y(i)=0.
        do 11 j=1,NTERM
          y(i)=y(i)+j*a(j)
11      continue
        y(i)=y(i)+SPREAD*gasdev(idum)
        sig(i)=SPREAD
12    continue
      do 13 i=1,NTERM
        ia(i)=1
13    continue
      call lfit(x,y,sig,NPT,a,ia,NTERM,covar,NTERM,chisq,funcs)
      write(*,'(/1x,t4,a,t22,a)') 'Parameter','Uncertainty'
      do 14 i=1,NTERM
        write(*,'(1x,t5,a,i1,a,f8.6,f11.6)') 'A(',i,') = ',
     *      a(i),sqrt(covar(i,i))
14    continue
      write(*,'(/3x,a,e12.6)') 'Chi-squared = ',chisq
      write(*,'(/3x,a)') 'Full covariance matrix'
      do 15 i=1,NTERM
        write(*,'(1x,6e12.2)') (covar(i,j),j=1,NTERM)
15    continue
      write(*,'(/1x,a)') 'press RETURN to continue...'
      read(*,*)
C     now check results of restricting fit parameters
      do 16 i=2,NTERM,2
        ia(i)=0
16    continue
      call lfit(x,y,sig,NPT,a,ia,NTERM,covar,NTERM,chisq,funcs)
      write(*,'(/1x,t4,a,t22,a)') 'Parameter','Uncertainty'
      do 17 i=1,NTERM
        write(*,'(1x,t5,a,i1,a,f8.6,f11.6)') 'A(',i,') = ',
     *      a(i),sqrt(covar(i,i))
17    continue
      write(*,'(/3x,a,e12.6)') 'Chi-squared = ',chisq
      write(*,'(/3x,a)') 'Full covariance matrix'
      do 18 i=1,NTERM
        write(*,'(1x,6e12.2)') (covar(i,j),j=1,NTERM)
18    continue
      END
```

```
      SUBROUTINE funcs(x,afunc,ma)
      INTEGER i,ma
      REAL x,afunc(ma)
      afunc(1)=1.
      afunc(2)=x
      do 11 i=3,ma
        afunc(i)=sin(i*x)
11    continue
      return
      END
```

covsrt is used in conjunction with lfit (and later with the routine svdfit) to redistribute the covariance matrix covar so that it represents the true order of coefficients, rather than the order in which they were fit. In sample routine xcovsrt an artificial 10×10 covariance matrix covar(i,j) is created, which is all zeros except for the upper left 5×5 section, for which the elements are covar(i,j)=i+j-1. Then two tests are performed:

1. By setting ia(i) = 1 for i = 1,...,10 and MFIT=5, we simulate a fit where the first 5 out of 10 parameters are fit. covar(i,j) should be returned unchanged.

2. By setting the even elements of the ia array to 1 and the odd elements to zero, we simulate a fit where parameters 2,4,6,8 and 10 have been fit. The elements of covar(i,j) should be spread so that alternate elements are zero.

```
      PROGRAM xcovsrt
C     driver for routine covsrt
      INTEGER MA,MFIT
      PARAMETER(MA=10,MFIT=5)
      INTEGER i,j,ia(MA)
      REAL covar(MA,MA)
      do 12 i=1,MA
        do 11 j=1,MA
          covar(i,j)=0.0
          if (i.le.MFIT .and. j.le.MFIT) then
            covar(i,j)=i+j-1
          endif
11      continue
12    continue
      write(*,'(//2x,a)') 'Original matrix'
      do 13 i=1,MA
        write(*,'(1x,10f4.1)') (covar(i,j),j=1,MA)
13    continue
      write(*,*) ' press RETURN to continue...'
      read(*,*)
      write(*,'(/2x,a)') 'Test #1 - Full Fitting'
      do 14 i=1,MA
        ia(i)=1
14    continue
      call covsrt(covar,MA,MA,ia,MA)
      do 15 i=1,MA
        write(*,'(1x,10f4.1)') (covar(i,j),j=1,MA)
15    continue
      write(*,*) ' press RETURN to continue...'
      read(*,*)
      write(*,'(/2x,a)') 'Test #2 - Spread'
```

```
        do 17 i=1,MA
          do 16 j=1,MA
            covar(i,j)=0.0
            if (i.le.MFIT .and. j.le.MFIT) then
              covar(i,j)=i+j-1
            endif
16        continue
17      continue
        do 18 i=1,MA,2
          ia(i)=0
18      continue
        call covsrt(covar,MA,MA,ia,MFIT)
        do 19 i=1,MA
          write(*,'(1x,10f4.1)') (covar(i,j),j=1,MA)
19      continue
        END
```

Routine `svdfit` is recommended in preference to `lfit` for performing linear least-squares fits. The sample program `xsvdfit` puts `svdfit` to work on the data generated according to

$$F(x) = 1 + 2x + 3x^2 + 4x^3 + 5x^4 + \text{Gaussian noise}.$$

This data is fit first to a five-term polynomial sum, and then to a five-term Legendre polynomial sum. In each case `sig(i)`, the measurement fluctuation in y, is taken to be constant. For the polynomial fit, the resulting coefficients should clearly have the values `a(i)` $\approx$ `i`. For Legendre polynomials the expected results are:

$$a(1) \approx 3.0$$

$$a(2) \approx 4.4$$

$$a(3) \approx 4.9$$

$$a(4) \approx 1.6$$

$$a(5) \approx 1.1$$

```
        PROGRAM xsvdfit
C       driver for routine svdfit
        INTEGER NPOL,NPT
        REAL SPREAD
        PARAMETER(NPT=100,SPREAD=0.02,NPOL=5)
        INTEGER i,idum,mp,np
        REAL chisq,gasdev
        REAL x(NPT),y(NPT),sig(NPT),a(NPOL),cvm(NPOL,NPOL)
        REAL u(NPT,NPOL),v(NPOL,NPOL),w(NPOL)
        EXTERNAL fpoly,fleg
C       polynomial fit
        idum=-911
        mp=NPT
        np=NPOL
        do 11 i=1,NPT
          x(i)=0.02*i
          y(i)=1.0+x(i)*(2.0+x(i)*(3.0+x(i)*(4.0+x(i)*5.0)))
          y(i)=y(i)*(1.0+SPREAD*gasdev(idum))
          sig(i)=y(i)*SPREAD
```

```
11    continue
      call svdfit(x,y,sig,NPT,a,NPOL,u,v,w,mp,np,chisq,fpoly)
      call svdvar(v,NPOL,np,w,cvm,NPOL)
      write(*,*) 'Polynomial fit:'
      do 12 i=1,NPOL
        write(*,'(1x,f12.6,a,f10.6)') a(i),' +-',sqrt(cvm(i,i))
12    continue
      write(*,'(1x,a,f12.6/)') 'Chi-squared',chisq
      call svdfit(x,y,sig,NPT,a,NPOL,u,v,w,mp,np,chisq,fleg)
      call svdvar(v,NPOL,np,w,cvm,NPOL)
      write(*,*) 'Legendre polynomial fit'
      do 13 i=1,NPOL
        write(*,'(1x,f12.6,a,f10.6)') a(i),' +-',sqrt(cvm(i,i))
13    continue
      write(*,'(1x,a,f12.6/)') 'Chi-squared',chisq
      END
```

svdvar is used with svdfit to evaluate the covariance matrix cvm of a fit with MA parameters. In program xsvdvar, we provide input vector w and array v for this routine via two data statements, and calculate the covariance matrix cvm determined from them. We have also done the calculation by hand and recorded the correct results in array tru for comparison.

```
      PROGRAM xsvdvar
C     driver for routine svdvar
      INTEGER NCVM,MA,MP
      PARAMETER(MP=6,MA=3,NCVM=MA)
      INTEGER i,j
      REAL v(MP,MP),w(MP),cvm(NCVM,NCVM),tru(MA,MA)
      DATA w/0.0,1.0,2.0,3.0,4.0,5.0/
      DATA v/1.0,2.0,3.0,4.0,5.0,6.0,1.0,2.0,3.0,4.0,5.0,6.0,
     *    1.0,2.0,3.0,4.0,5.0,6.0,1.0,2.0,3.0,4.0,5.0,6.0,
     *    1.0,2.0,3.0,4.0,5.0,6.0,1.0,2.0,3.0,4.0,5.0,6.0/
      DATA tru/1.25,2.5,3.75,2.5,5.0,7.5,3.75,7.5,11.25/
      write(*,'(/1x,a)') 'Matrix V'
      do 11 i=1,MP
        write(*,'(1x,6f12.6)') (v(i,j),j=1,MP)
11    continue
      write(*,'(/1x,a)') 'Vector W'
      write(*,'(1x,6f12.6)') (w(i),i=1,MP)
      call svdvar(v,MA,MP,w,cvm,NCVM)
      write(*,'(/1x,a)') 'Covariance matrix from SVDVAR'
      do 12 i=1,MA
        write(*,'(1x,3f12.6)') (cvm(i,j),j=1,MA)
12    continue
      write(*,'(/1x,a)') 'Expected covariance matrix'
      do 13 i=1,MA
        write(*,'(1x,3f12.6)') (tru(i,j),j=1,MA)
13    continue
      END
```

Routines fpoly and fleg were used above with sample program xsvdfit to generate the powers of x and the Legendre polynomials, respectively. In the case of fpoly, sample program xfpoly below is used to list the powers of x generated by fpoly so that they may be checked "by eye". For fleg, the generated polynomials in program xfleg are compared to values from routine plgndr.

```
            PROGRAM xfpoly
C           driver for routine fpoly
            INTEGER NPOLY,NVAL
            REAL DX
            PARAMETER(NVAL=15,DX=0.1,NPOLY=5)
            INTEGER i,j
            REAL x,afunc(NPOLY)
            write(*,'(/1x,t29,a)') 'Powers of X'
            write(*,'(/1x,t9,a,t17,a,t27,a,t37,a,t47,a,t57,a)') 'X','X**0',
        *       'X**1','X**2','X**3','X**4'
            do 11 i=1,NVAL
              x=i*DX
              call fpoly(x,afunc,NPOLY)
              write(*,'(1x,6f10.4)') x,(afunc(j),j=1,NPOLY)
11          continue
            END

            PROGRAM xfleg
C           driver for routine fleg
            INTEGER NPOLY,NVAL
            REAL DX
            PARAMETER(NVAL=5,DX=0.2,NPOLY=5)
            INTEGER i,j
            REAL x,afunc(NPOLY),plgndr
            write(*,'(/1x,t25,a)') 'Legendre Polynomials'
            write(*,'(/1x,t8,a,t18,a,t28,a,t38,a,t48,a)')
        *       'N=1','N=2','N=3','N=4','N=5'
            do 11 i=1,NVAL
              x=i*DX
              call fleg(x,afunc,NPOLY)
              write(*,'(1x,a,f6.2)') 'X =',x
              write(*,'(1x,5f10.4,a)') (afunc(j),j=1,NPOLY),' routine FLEG'
              write(*,'(1x,5f10.4,a/)') (plgndr(j-1,0,x),j=1,NPOLY),
        *          ' routine PLGNDR'
11          continue
            END
```

`mrqmin` is used along with `mrqcof` to perform nonlinear least-squares fits with the Levenberg-Marquardt method. The artificial data used to try it in sample program `xmrqmin` is computed as the sum of two Gaussians plus noise:

$$y(i) = a(1) \exp\{-[(x(i) - a(2))/a(3)]^2\}$$
$$+ a(4) \exp\{-[(x(i) - a(5))/a(6)]^2\} + \text{noise}.$$

The $a(i)$ are set up in a `DATA` statement, as are the initial guesses `gues(i)` for these parameters to be used in initiating the fit. Also initialized for the fit are `ia(i)=1` for $i=1,\ldots,\text{mfit}$ to specify that all six of the parameters are to be fit. On the first call to `mrqmin`, `alamda=-1` to initialize. Then a loop is entered in which `mrqmin` is iterated while monitoring successive values of chi-squared `chisq`. When `chisq` changes by less than 0.1 on two consecutive iterations, the fit is considered complete, and `mrqmin` is called one final time with `alamda=0.0` so that array `covar` will return the covariance matrix. Uncertainties are derived from the square roots of the diagonal elements of `covar`. Expected results for the parameters are, of course, the values used to generate the "data" in the first place. As a test of the `ia` feature, the fit is redone holding fixed `a(2)=2` and `a(5)=5`.

```
      PROGRAM xmrqmin
C     driver for routine mrqmin
      INTEGER NPT,MA
      REAL SPREAD
      PARAMETER(NPT=100,MA=6,SPREAD=0.001)
      INTEGER i,ia(MA),idum,iter,itst,j,k,mfit
      REAL alamda,chisq,fgauss,gasdev,ochisq,x(NPT),y(NPT),sig(NPT),
     *      a(MA),covar(MA,MA),alpha(MA,MA),gues(MA)
      EXTERNAL fgauss
      DATA a/5.0,2.0,3.0,2.0,5.0,3.0/
      DATA gues/4.5,2.2,2.8,2.5,4.9,2.8/
      idum=-911
C     first try a sum of two Gaussians
      do 12 i=1,NPT
        x(i)=0.1*i
        y(i)=0.0
        do 11 j=1,MA,3
          y(i)=y(i)+a(j)*exp(-((x(i)-a(j+1))/a(j+2))**2)
11      continue
        y(i)=y(i)*(1.0+SPREAD*gasdev(idum))
        sig(i)=SPREAD*y(i)
12    continue
      mfit=MA
      do 13 i=1,mfit
        ia(i)=1
13    continue
      do 14 i=1,MA
        a(i)=gues(i)
14    continue
      do 16 iter=1,2
        alamda=-1
        call mrqmin(x,y,sig,NPT,a,ia,MA,covar,alpha,
     *        MA,chisq,fgauss,alamda)
        k=1
        itst=0
1       write(*,'(/1x,a,i2,t18,a,f10.4,t43,a,e9.2)') 'Iteration #',k,
     *        'Chi-squared:',chisq,'ALAMDA:',alamda
        write(*,'(1x,t5,a,t13,a,t21,a,t29,a,t37,a,t45,a)') 'A(1)',
     *        'A(2)','A(3)','A(4)','A(5)','A(6)'
        write(*,'(1x,6f8.4)') (a(i),i=1,6)
        k=k+1
        ochisq=chisq
        call mrqmin(x,y,sig,NPT,a,ia,MA,covar,alpha,
     *        MA,chisq,fgauss,alamda)
        if (chisq.gt.ochisq) then
          itst=0
        else if (abs(ochisq-chisq).lt.0.1) then
          itst=itst+1
        endif
        if (itst.lt.4) then
          goto 1
        endif
        alamda=0.0
        call mrqmin(x,y,sig,NPT,a,ia,MA,covar,alpha,
     *        MA,chisq,fgauss,alamda)
        write(*,*) 'Uncertainties:'
        write(*,'(1x,6f8.4/)') (sqrt(covar(i,i)),i=1,6)
```

```
        write(*,'(1x,a)') 'Expected results:'
        write(*,'(1x,f7.2,5f8.2/)') 5.0,2.0,3.0,2.0,5.0,3.0
        if (iter.eq.1) then
          write(*,*) 'press return to continue with constraint'
          read(*,*)
          write(*,*) 'Holding a(2) and a(5) constant'
          do 15 j=1,MA
            a(j)=a(j)+.1
15        continue
          a(2)=2.0
          ia(2)=0
          a(5)=5.0
          ia(5)=0
        endif
16      continue
        END
```

The nonlinear least-squares fit makes use of a vector β_k (the gradient of χ^2 in parameter-space) and α_{kl} (the Hessian of χ^2 in the same space). These quantities are produced by mrqcof, as demonstrated by sample program xmrqcof. The function is a sum of two Gaussians with noise added (the same function as in xmrqmin) and it is used twice. In the first call, ia(i)=1 and mfit=6 so all six parameters are used. In the second call, ia(i)=0 for i=1,2,3 and mfit=3 so the first three parameters are fixed and the last three, a(4)...a(6) are fit. Here is the output you should get:

```
matrix alpha
    137.2021     60.6572    148.2204    147.7027    -12.6093     90.9625
     60.6572    500.8828    366.9686    448.3021    287.4484    257.7019
    148.2204    366.9686    583.6553    659.6396    553.0109    612.6782
    147.7027    448.3021    659.6396    971.0058   1131.7703   1410.9384
    -12.6093    287.4484    553.0109   1131.7703   1911.5937   2409.3491
     90.9625    257.7019    612.6782   1410.9384   2409.3491   3341.0012
vector beta
    -24.7629   -118.2998    -76.5074   -124.1268   -104.8776   -145.4019
Chi-squared:    147.9271
matrix alpha
    971.0058   1131.7703   1410.9384
   1131.7703   1911.5937   2409.3491
   1410.9384   2409.3491   3341.0012
vector beta
   -124.1268   -104.8776   -145.4019
Chi-squared:    147.9271
```

```
        PROGRAM xmrqcof
C       driver for routine mrqcof
        INTEGER NPT,MA
        REAL SPREAD
        PARAMETER(NPT=100,MA=6,SPREAD=0.1)
        INTEGER i,idum,j,ia(MA),mfit
        REAL chisq,gasdev,x(NPT),y(NPT),sig(NPT),a(MA),
     *      alpha(MA,MA),beta(MA),gues(MA)
        EXTERNAL fgauss
        DATA a/5.0,2.0,3.0,2.0,5.0,3.0/
        DATA gues/4.9,2.1,2.9,2.1,4.9,3.1/
        idum=-911
C       first try sum of two gaussians
```

```
      do 12 i=1,NPT
        x(i)=0.1*i
        y(i)=0.0
        do 11 j=1,4,3
          y(i)=y(i)+a(j)*exp(-((x(i)-a(j+1))/a(j+2))**2)
11      continue
        y(i)=y(i)*(1.0+SPREAD*gasdev(idum))
        sig(i)=SPREAD*y(i)
12    continue
      mfit=MA
      do 13 i=1,mfit
        ia(i)=1
13    continue
      do 14 i=1,mfit
        a(i)=gues(i)
14    continue
      call mrqcof(x,y,sig,NPT,a,ia,MA,alpha,beta,MA,chisq,fgauss)
      write(*,'(/1x,a)') 'matrix alpha'
      do 15 i=1,MA
        write(*,'(1x,6f12.4)') (alpha(i,j),j=1,MA)
15    continue
      write(*,'(1x,a)') 'vector beta'
      write(*,'(1x,6f12.4)') (beta(i),i=1,MA)
      write(*,'(1x,a,f12.4/)') 'Chi-squared:',chisq
C     next fix one line and improve the other
      mfit=3
      do 16 i=1,mfit
        ia(i)=0
16    continue
      do 17 i=1,MA
        a(i)=gues(i)
17    continue
      call mrqcof(x,y,sig,NPT,a,ia,MA,alpha,beta,MA,chisq,fgauss)
      write(*,'(1x,a)') 'matrix alpha'
      do 18 i=1,mfit
        write(*,'(1x,6f12.4)') (alpha(i,j),j=1,mfit)
18    continue
      write(*,'(1x,a)') 'vector beta'
      write(*,'(1x,6f12.4)') (beta(i),i=1,mfit)
      write(*,'(1x,a,f12.4/)') 'Chi-squared:',chisq
      END
```

fgauss is an example of the type of subroutine that must be supplied to mrqfit in order to fit a user-defined function, in this case the sum of Gaussians. fgauss calculates both the function, and its derivative with respect to each adjustable parameter in a fairly compact fashion. The sample program xfgauss calculates the same quantities in a more pedantic fashion, just to be sure we got everything right.

```
      PROGRAM xfgauss
C     driver for routine fgauss
      INTEGER NA,NLIN,NPT
      PARAMETER(NPT=3,NLIN=2,NA=3*NLIN)
      INTEGER i,j
      REAL e1,e2,f,x,y,a(NA),dyda(NA),df(NA)
      DATA a/3.0,0.2,0.5,1.0,0.7,0.3/
      write(*,'(/1x,t6,a,t14,a,t19,a,t27,a,t35,a,t43,a,t51,a,t59,a)')
     *     'X','Y','DYDA1','DYDA2','DYDA3','DYDA4','DYDA5','DYDA6'
```

```
      do 11 i=1,NPT
        x=0.3*i
        call fgauss(x,a,y,dyda,NA)
        e1=exp(-((x-a(2))/a(3))**2)
        e2=exp(-((x-a(5))/a(6))**2)
        f=a(1)*e1+a(4)*e2
        df(1)=e1
        df(4)=e2
        df(2)=a(1)*e1*2.0*(x-a(2))/(a(3)**2)
        df(5)=a(4)*e2*2.0*(x-a(5))/(a(6)**2)
        df(3)=a(1)*e1*2.0*((x-a(2))**2)/(a(3)**3)
        df(6)=a(4)*e2*2.0*((x-a(5))**2)/(a(6)**3)
        write(*,'(1x,a/,8f8.4)') 'from FGAUSS',x,y,(dyda(j),j=1,6)
        write(*,'(1x,a/,8f8.4/)') 'independent calc.',x,f,(df(j),j=1,6)
11    continue
      END
```

medfit is a subroutine illustrating a more "robust" way of fitting. It performs a fit of data to a straight line, but instead of using the least-squares criterion for figuring the merit of a fit, it uses the least-absolute-deviation. For comparison, sample routine xmedfit fits lines to a noisy linear data set, using first the least-squares routine fit, and then the least-absolute-deviation routine medfit. You may be interested to see if you can figure out what mean value of absolute deviation you expect for data with gaussian noise of amplitude SPREAD.

```
      PROGRAM xmedfit
C     driver for routine medfit
      INTEGER NPT
      REAL SPREAD
      PARAMETER(NPT=100,SPREAD=0.1)
      INTEGER i,idum,mwt
      REAL a,abdev,b,chi2,gasdev,q,siga,sigb,x(NPT),y(NPT),sig(NPT)
      idum=-1984
      do 11 i=1,NPT
        x(i)=0.1*i
        y(i)=-2.0*x(i)+1.0+SPREAD*gasdev(idum)
        sig(i)=SPREAD
11    continue
      mwt=1
      call fit(x,y,NPT,sig,mwt,a,b,siga,sigb,chi2,q)
      write(*,'(/1x,a)') 'According to routine FIT the result is:'
      write(*,'(1x,t5,a,f8.4,t20,a,f8.4)') 'A = ',a,'Uncertainty: ',
     *     siga
      write(*,'(1x,t5,a,f8.4,t20,a,f8.4)') 'B = ',b,'Uncertainty: ',
     *     sigb
      write(*,'(1x,t5,a,f8.4,a,i4,a)') 'Chi-squared: ',chi2,
     *     ' for ',NPT,' points'
      write(*,'(1x,t5,a,f8.4)') 'Goodness-of-fit: ',q
      write(*,'(/1x,a)') 'According to routine MEDFIT the result is:'
      call medfit(x,y,NPT,a,b,abdev)
      write(*,'(1x,t5,a,f8.4)') 'A = ',a
      write(*,'(1x,t5,a,f8.4)') 'B = ',b
      write(*,'(1x,t5,a,f8.4)') 'Absolute deviation (per DATA point): '
     *     ,abdev
      write(*,'(1x,t5,a,f8.4,a)') '(note: Gaussian SPREAD is',SPREAD,')'
      END
```

rofunc is an auxiliary function for medfit. It evaluates the quantity

$$\sum_{i=1}^{N} x_i \operatorname{sgn}(y_i - a - bx_i)$$

given arrays x_i and y_i. Data are communicated to and from rofunc primarily through the common block arrays, but the value of the sum above is returned as the value of rofunc(b). abdev is the summed absolute deviation, and aa (listed below as A) is given the value that minimizes abdev. Our results for these quantities are:

B	A	ROFUNC	ABDEV
-2.10	1.51	249.80	25.17
-2.08	1.40	247.40	20.20
-2.06	1.30	247.40	15.25
-2.04	1.21	231.20	10.36
-2.02	1.11	193.60	5.92
-2.00	1.00	-15.60	3.75
-1.98	0.89	-199.20	6.00
-1.96	0.80	-234.40	10.40
-1.94	0.71	-247.20	15.25
-1.92	0.60	-247.20	20.19
-1.90	0.51	-247.20	25.14

```
      PROGRAM xrofunc
C     driver for routine rofunc
      INTEGER NMAX
      REAL SPREAD
      PARAMETER(NMAX=1000,SPREAD=0.05)
      INTEGER i,idum,npt
      REAL aa,abdev,x(NMAX),y(NMAX),arr(NMAX),b,rf,gasdev,rofunc
      COMMON /arrays/ x,y,arr,aa,abdev,npt
      idum=-11
      npt=100
      do 11 i=1,npt
        x(i)=0.1*i
        y(i)=-2.0*x(i)+1.0+SPREAD*gasdev(idum)
11    continue
      write(*,'(/1x,t10,a,t20,a,t26,a,t37,a/)') 'B','A','ROFUNC','ABDEV'
      do 12 i=-5,5
        b=-2.0+0.02*i
        rf=rofunc(b)
        write(*,'(1x,4f10.2)') b,aa,rofunc(b),abdev
12    continue
      END
```

Chapter 16: Ordinary Differential Equations

*Chapter 16 of Numerical Recipes deals with the integration of ordinary differential equations, restricting its attention specifically to initial-value problems. Three practical methods are introduced: 1) Runge-Kutta methods (*rk4, rkdumb, rkqs,* and* odeint*), 2) Richardson extrapolation and the Bulirsch-Stoer method (*bsstep, mmid, stoerm, rzextr, pzextr*), 3) predictor-corrector methods. In general, for applications not demanding high precision, and where convenience is paramount, Runge-Kutta with adaptive step-size control is recommended. For higher precision applications, the Bulirsch-Stoer method dominates. The predictor-corrector methods are covered because of their history of widespread use.*

Stiff differential equations require special methods. Routine stiff *uses a generalization of Runge-Kutta, known as a Rosenbrock method. The semi-implicit midpoint extrapolation method is a Bulirsch-Stoer method designed for stiff equations, and implemented in* stifbs.

$$\star \quad \star \quad \star \quad \star$$

Routine rk4 advances the solution vector y(n) of a set of ordinary differential equations over a single small interval h in x using the fourth-order Runge-Kutta method. The operation is shown by sample program xrk4 for an array of four variables y(1),...,y(4). The first-order differential equations satisfied by these variables are specified by the accompanying routine derivs, and are simply the equations describing the first four Bessel functions $J_0(x),\ldots,J_3(x)$. The y's are initialized to the values of these functions at $x = 1.0$. Note that the values of dydx are also initialized at $x = 1.0$, because rk4 uses the values of dydx before its first call to derivs. The reason for this is discussed in the text. The sample program calls rk4 with h (the step-size) set to various values from 0.2 to 1.0, so that you can see how well rk4 can do even with quite sizeable steps.

```
      PROGRAM xrk4
C     driver for routine rk4
      INTEGER N
      PARAMETER(N=4)
      INTEGER i,j
      REAL bessj,bessj0,bessj1
      REAL h,x,y(N),dydx(N),yout(N)
      EXTERNAL derivs
      x=1.0
      y(1)=bessj0(x)
      y(2)=bessj1(x)
      y(3)=bessj(2,x)
      y(4)=bessj(3,x)
```

```
      call derivs(x,y,dydx)
      write(*,'(/1x,a,t19,a,t31,a,t43,a,t55,a)')
     *     'Bessel Function:','J0','J1','J3','J4'
      do 11 i=1,5
        h=0.2*i
        call rk4(y,dydx,N,x,h,yout,derivs)
        write(*,'(/1x,a,f6.2)') 'For a step size of:',h
        write(*,'(1x,a10,4f12.6)') 'RK4:',(yout(j),j=1,4)
        write(*,'(1x,a10,4f12.6)') 'Actual:',bessj0(x+h),
     *       bessj1(x+h),bessj(2,x+h),bessj(3,x+h)
11    continue
      END

      SUBROUTINE derivs(x,y,dydx)
      REAL x,y(*),dydx(*)
      dydx(1)=-y(2)
      dydx(2)=y(1)-(1.0/x)*y(2)
      dydx(3)=y(2)-(2.0/x)*y(3)
      dydx(4)=y(3)-(3.0/x)*y(4)
      return
      END
```

rkdumb is an extension of rk4 that allows you to integrate over larger intervals. It is "dumb" in the sense that it has no adaptive step-size determination, and no code to estimate errors. Sample program xrkdumb works with the same functions and derivatives as the previous program, but integrates from x1=1.0 to x2=20.0, breaking the interval into NSTEP=150 equal steps. The variables vstart(1),...,vstart(4) which become the starting values of the y's, are initialized as before, but their derivatives this time are not initialized; rkdumb takes care of that. This time only the results for the fourth variable $J_3(x)$ are listed, and only every tenth value is given. The values are passed in the named common block path.

```
      PROGRAM xrkdumb
C     driver for routine rkdumb
      INTEGER NSTEP,NVAR
      PARAMETER(NVAR=4,NSTEP=150)
      INTEGER i,j
      REAL bessj,bessj0,bessj1
      REAL x(200),x1,x2,y(50,200),vstart(NVAR)
      COMMON /path/ x,y
      EXTERNAL derivs
      x1=1.0
      vstart(1)=bessj0(x1)
      vstart(2)=bessj1(x1)
      vstart(3)=bessj(2,x1)
      vstart(4)=bessj(3,x1)
      x2=20.0
      call rkdumb(vstart,NVAR,x1,x2,NSTEP,derivs)
      write(*,'(/1x,t9,a,t17,a,t31,a/)') 'X','Integrated','BESSJ3'
      do 11 i=1,(NSTEP/10)
        j=10*i
        write(*,'(1x,f10.4,2x,2f12.6)') x(j),y(4,j),bessj(3,x(j))
11    continue
      END

      SUBROUTINE derivs(x,y,dydx)
```

```
REAL x,y(*),dydx(*)
dydx(1)=-y(2)
dydx(2)=y(1)-(1.0/x)*y(2)
dydx(3)=y(2)-(2.0/x)*y(3)
dydx(4)=y(3)-(3.0/x)*y(4)
return
END
```

rkqs performs a single step of fifth-order Runge-Kutta integration, this time with monitoring of local truncation error and corresponding step-size adjustment. Its sample program xrkqs is similar to that for routine rk4, using four Bessel functions as the example, and starting the integration at $x = 1.0$. However, on each pass a value is set for eps, the desired accuracy, and the trial value htry for the interval size is set to 0.6. For the first few passes, eps is not too demanding and htry may be perfectly adequate. As eps becomes smaller, the routine will be forced to diminish h and return smaller values of hdid and hnext. Our results (in single precision) are:

eps	htry	hdid	hnext
0.3679E+00	0.60	0.600000	3.000000
0.1353E+00	0.60	0.600000	2.542160
0.4979E-01	0.60	0.600000	2.081344
0.1832E-01	0.60	0.600000	1.704060
0.6738E-02	0.60	0.600000	1.395167
0.2479E-02	0.60	0.600000	1.142266
0.9119E-03	0.60	0.600000	0.935208
0.3355E-03	0.60	0.600000	0.765684
0.1234E-03	0.60	0.600000	0.626889
0.4540E-04	0.60	0.506776	0.513415
0.1670E-04	0.60	0.394678	0.421235
0.6144E-05	0.60	0.307375	0.346008
0.2260E-05	0.60	0.239384	0.284364
0.8315E-06	0.60	0.186432	0.233721
0.3059E-06	0.60	0.145194	0.192100

```
      PROGRAM xrkqs
C     driver for routine rkqs
      INTEGER N
      PARAMETER(N=4)
      INTEGER i,j
      REAL bessj,bessj0,bessj1
      REAL eps,hdid,hnext,htry,x,y(N),dydx(N),dysav(N),ysav(N),yscal(N)
      EXTERNAL derivs
      x=1.0
      ysav(1)=bessj0(x)
      ysav(2)=bessj1(x)
      ysav(3)=bessj(2,x)
      ysav(4)=bessj(3,x)
      call derivs(x,ysav,dysav)
      do 11 i=1,N
        yscal(i)=1.0
11    continue
      htry=0.6
      write(*,'(/1x,t8,a,t19,a,t31,a,t43,a)')
     *     'eps','htry','hdid','hnext'
      do 13 i=1,15
        eps=exp(-float(i))
```

```
      x=1.0
      do 12 j=1,N
        y(j)=ysav(j)
        dydx(j)=dysav(j)
12      continue
      call rkqs(y,dydx,n,x,htry,eps,yscal,hdid,hnext,derivs)
      write(*,'(2x,e12.4,f8.2,2x,2f12.6)') eps,htry,hdid,hnext
13    continue
      END

      SUBROUTINE derivs(x,y,dydx)
      REAL x,y(*),dydx(*)
      dydx(1)=-y(2)
      dydx(2)=y(1)-(1.0/x)*y(2)
      dydx(3)=y(2)-(2.0/x)*y(3)
      dydx(4)=y(3)-(3.0/x)*y(4)
      return
      END
```

The full driver routine for rkqs, which provides Runge-Kutta integration over large intervals with adaptive step-size control, is odeint. It plays the same role for rkqs that rkdumb plays for rk4, and like rkdumb it stores intermediate results in a common block named path. Integration is performed on four Bessel functions from x1=1.0 to x2=10.0, with an accuracy eps=1.0e-4. Independent of the values of step-size actually used by odeint, intermediate values will be recorded only at intervals greater than dxsav. The sample program returns values of $J_3(x)$ for checking against actual values produced by bessj. It also records the number of function evaluations used, how many steps were successful, and how many were "bad". Bad steps are redone, and indicate no extra loss in accuracy. At the same time, they do represent a loss in efficiency, so that an excessive number of bad steps should initiate an investigation.

```
      PROGRAM xodeint
C     driver for routine odeint
      INTEGER KMAXX,NMAX,NVAR
      PARAMETER (KMAXX=200,NMAX=50,NVAR=4)
      INTEGER i,kmax,kount,nbad,nok,nrhs
      REAL bessj,bessj0,bessj1
      REAL dxsav,eps,h1,hmin,x1,x2,x,y,ystart(NVAR)
      COMMON /path/ kmax,kount,dxsav,x(KMAXX),y(NMAX,KMAXX)
      COMMON nrhs
      EXTERNAL derivs,rkqs
      nrhs=0
      x1=1.0
      x2=10.0
      ystart(1)=bessj0(x1)
      ystart(2)=bessj1(x1)
      ystart(3)=bessj(2,x1)
      ystart(4)=bessj(3,x1)
      eps=1.0e-4
      h1=.1
      hmin=0.0
      kmax=100
      dxsav=(x2-x1)/20.0
      call odeint(ystart,NVAR,x1,x2,eps,h1,hmin,nok,nbad,derivs,rkqs)
```

```
      write(*,'(/1x,a,t30,i3)') 'Successful steps:',nok
      write(*,'(1x,a,t30,i3)') 'Bad steps:',nbad
      write(*,'(1x,a,t30,i3)') 'Function evaluations:',nrhs
      write(*,'(1x,a,t30,i3)') 'Stored intermediate values:',kount
      write(*,'(/1x,t9,a,t20,a,t33,a)') 'X','Integral','BESSJ(3,X)'
      do 11 i=1,kount
        write(*,'(1x,f10.4,2x,2f14.6)') x(i),y(4,i),bessj(3,x(i))
11    continue
      END

      SUBROUTINE derivs(x,y,dydx)
      INTEGER nrhs
      REAL x,y(*),dydx(*)
      COMMON nrhs
      nrhs=nrhs+1
      dydx(1)=-y(2)
      dydx(2)=y(1)-(1.0/x)*y(2)
      dydx(3)=y(2)-(2.0/x)*y(3)
      dydx(4)=y(3)-(3.0/x)*y(4)
      return
      END
```

The modified midpoint routine mmid is presented in *Numerical Recipes* primarily as a component of the more powerful Bulirsch-Stoer routine. It integrates variables over an interval htot through a sequence of much smaller steps. Sample routine xmmid takes the number of subintervals i to be $5, 10, 15, \ldots, 50$ so that we can witness any improvements in accuracy that may occur. The values of the four Bessel functions are compared with the results of the integrations.

```
      PROGRAM xmmid
C     driver for routine mmid
      INTEGER NVAR
      REAL HTOT,X1
      PARAMETER(NVAR=4,X1=1.0,HTOT=0.5)
      INTEGER i
      REAL bessj,bessj0,bessj1
      REAL b1,b2,b3,b4,xf,y(NVAR),yout(NVAR),dydx(NVAR)
      EXTERNAL derivs
      y(1)=bessj0(X1)
      y(2)=bessj1(X1)
      y(3)=bessj(2,X1)
      y(4)=bessj(3,X1)
      call derivs(X1,y,dydx)
      xf=X1+HTOT
      b1=bessj0(xf)
      b2=bessj1(xf)
      b3=bessj(2,xf)
      b4=bessj(3,xf)
      write(*,'(1x,a/)') 'First four Bessel functions'
      do 11 i=5,50,5
        call mmid(y,dydx,NVAR,X1,HTOT,i,yout,derivs)
        write(*,'(1x,a,f6.4,a,f6.4,a,i2,a)') 'X = ',X1,
     *        ' to ',X1+HTOT,' in ',i,' steps'
        write(*,'(1x,t5,a,t20,a)') 'Integration','BESSJ'
        write(*,'(1x,2f12.6)') yout(1),b1
        write(*,'(1x,2f12.6)') yout(2),b2
```

```
      write(*,'(1x,2f12.6)') yout(3),b3
      write(*,'(1x,2f12.6)') yout(4),b4
      write(*,'(/1x,a)') 'press RETURN to continue...'
      read(*,*)
11    continue
      END

      SUBROUTINE derivs(x,y,dydx)
      REAL x,y(*),dydx(*)
      dydx(1)=-y(2)
      dydx(2)=y(1)-(1.0/x)*y(2)
      dydx(3)=y(2)-(2.0/x)*y(3)
      dydx(4)=y(3)-(3.0/x)*y(4)
      END
```

The Bulirsch-Stoer method, carried out by routine `bsstep`, is the integrator of choice for higher accuracy calculations on smooth functions. An interval h is broken into finer and finer steps, and the results of integration are extrapolated to zero step-size. The extrapolation is via a polynomial with `pzextr`. `bsstep` monitors local truncation error and adjusts the step-size appropriately, to keep errors below `eps`. From an external point of view, `bsstep` operates exactly as does `rkqs`: it has the same arguments and in the same order. Consequently it can be used in place of `rkqs` in routine `odeint`, allowing more efficient integration over large regions of x. For this reason, the sample program `xbsstep` is of the same form used to demonstrate `rkqs`. You will notice that `bsstep` integrates the test problem in far fewer steps than `rkqs`, which is typical. Because of the low accuracy requirement, however, `rkqs` makes fewer function evaluations.

```
      PROGRAM xbsstep
C     driver for routine bsstep
      INTEGER KMAXX,NMAX,NVAR
      PARAMETER (KMAXX=200,NMAX=50,NVAR=4)
      INTEGER i,kmax,kount,nbad,nok,nrhs
      REAL bessj,bessj0,bessj1
      REAL dxsav,eps,h1,hmin,x1,x2,x,y,ystart(NVAR)
      COMMON /path/ kmax,kount,dxsav,x(KMAXX),y(NMAX,KMAXX)
      COMMON nrhs
      EXTERNAL derivs,bsstep
      nrhs=0
      x1=1.0
      x2=10.0
      ystart(1)=bessj0(x1)
      ystart(2)=bessj1(x1)
      ystart(3)=bessj(2,x1)
      ystart(4)=bessj(3,x1)
      eps=1.0e-4
      h1=.1
      hmin=0.0
      kmax=100
      dxsav=(x2-x1)/20.0
      call odeint(ystart,NVAR,x1,x2,eps,h1,hmin,nok,nbad,derivs,bsstep)
      write(*,'(/1x,a,t30,i3)') 'Successful steps:',nok
      write(*,'(1x,a,t30,i3)') 'Bad steps:',nbad
      write(*,'(1x,a,t30,i3)') 'Function evaluations:',nrhs
      write(*,'(1x,a,t30,i3)') 'Stored intermediate values:',kount
```

```
      write(*,'(/1x,t9,a,t20,a,t33,a)')') 'X','Integral','BESSJ(3,X)'
      do 11 i=1,kount
        write(*,'(1x,f10.4,2x,2f14.6)')') x(i),y(4,i),bessj(3,x(i))
11    continue
      END

      SUBROUTINE derivs(x,y,dydx)
      INTEGER nrhs
      REAL x,y(*),dydx(*)
      COMMON nrhs
      nrhs=nrhs+1
      dydx(1)=-y(2)
      dydx(2)=y(1)-(1.0/x)*y(2)
      dydx(3)=y(2)-(2.0/x)*y(3)
      dydx(4)=y(3)-(3.0/x)*y(4)
      return
      END
```

pzextr performs a polynomial extrapolation for bsstep. It takes a sequence of interval lengths and corresponding integrated values, and extrapolates to the value the integral would have if the interval length were zero. Sample routine xpzextr works with a known function

$$F_n = \frac{1 - x + x^3}{(x + 1)^n} \qquad n = 1, .., 4$$

We extrapolate the vector yest $= (F_1, F_2, F_3, F_4)$ given a sequence of ten values. The ten values are labelled iest=1,..., 10 and are evaluated at xest=1.0/iest. A call to pzextr produces extrapolated values yz, and estimated errors dy, and compares to the true values $(1.0, 1.0, 1.0, 1.0)$ at xest=0.0.

```
      PROGRAM xpzextr
C     driver for routine pzextr
      INTEGER NV
      PARAMETER(NV=4)
      INTEGER i,iest,j
      REAL dum,xest,yest(NV),yz(NV),dy(NV)
      do 12 i=1,10
        iest=i
        xest=1.0/float(i)
        dum=1.0-xest+xest*xest*xest
        do 11 j=1,NV
          dum=dum/(xest+1.0)
          yest(j)=dum
11      continue
        call pzextr(iest,xest,yest,yz,dy,NV)
        write(*,'(/1x,a,i2)') 'I = ',i
        write(*,'(1x,a,4f12.6)') 'Extrap. function:',(yz(j),j=1,NV)
        write(*,'(1x,a,4f12.6)') 'Estimated error: ',(dy(j),j=1,NV)
12    continue
      write(*,'(/1x,a,4f12.6)') 'Actual values:    ',1.0,1.0,1.0,1.0
      END
```

rzextr performs a diagonal rational function extrapolation for bsstep. It is usually a less efficient standby for rzextr, to be used primarily when some problem crops up with the extrapolation. The sample program xrzextr is identical to that for pzextr.

```
      PROGRAM xrzextr
C     driver for routine rzextr
      INTEGER NV
      PARAMETER(NV=4)
      INTEGER i,iest,j
      REAL dum,xest,yest(NV),yz(NV),dy(NV)
      do 12 i=1,10
        iest=i
        xest=1.0/float(i)
        dum=1.0-xest+xest**3
        do 11 j=1,NV
          dum=dum/(xest+1.0)
          yest(j)=dum
11      continue
        call rzextr(iest,xest,yest,yz,dy,NV)
        write(*,'(/1x,a,i2,a,f8.4)') 'IEST = ',i,'   XEST =',xest
        write(*,'(1x,a,4f12.6)') 'Extrap. Function: ',(yz(j),j=1,NV)
        write(*,'(1x,a,4f12.6)') 'Estimated Error:  ',(dy(j),j=1,NV)
12    continue
      write(*,'(/1x,a,4f12.6)') 'Actual Values:    ',1.0,1.0,1.0,1.0
      END
```

Stoermer's method is a special-purpose integration method for second-order ODEs in which the first derivative term is absent. It replaces `mmid` in the Bulirsch-Stoer method. Sample program `xstoerm` tests it out on the system

$$y_1'' = x - y_1$$
$$y_2'' = x^2 - y_2$$

The system is integrated from $x = 0$ to $x = \pi/2$ with initial conditions

$$y_1 = 0, \quad y_2 = -1, \quad y_1' = 2, \quad y_2' = 0$$

The program compares the result of integrating with varying number of substeps to the exact solution

$$y_1 = x + \sin x$$
$$y_2 = x^2 + \cos x - 2$$

Of course, in real life you would use `stoerm` as a replacement for `mmid` in a modified version of `bsstep`, as described in *Numerical Recipes*.

```
      PROGRAM xstoerm
C     driver for routine stoerm
      INTEGER NVAR
      REAL HTOT,X1
      PARAMETER(NVAR=4,X1=0.,HTOT=1.570796)
      INTEGER i
      REAL a1,a2,x,xf,y(NVAR),yout(NVAR),d2y(NVAR),d2y1,d2y2
      EXTERNAL derivs
      d2y1(x)=x+sin(x)
      d2y2(x)=x**2+cos(x)-2
      y(1)=0.
      y(2)=-1.
      y(3)=2.
      y(4)=0.
      call derivs(X1,y,d2y)
```

```
      xf=X1+HTOT
      a1=d2y1(xf)
      a2=d2y2(xf)
      write(*,'(1x,a/)') 'Stoermer''s Rule:'
      do 11 i=5,45,10
        call stoerm(y,d2y,NVAR,X1,HTOT,i,yout,derivs)
        write(*,'(1x,a,f6.4,a,f6.4,a,i2,a)') 'X = ',X1,
     *       ' to ',X1+HTOT,' in ',i,' steps'
        write(*,'(1x,t5,a,t20,a)') 'Integration','Answer'
        write(*,'(1x,2f12.6)') yout(1),a1
        write(*,'(1x,2f12.6)') yout(2),a2
11    continue
      END

      SUBROUTINE derivs(x,y,d2y)
      REAL x,y(*),d2y(*)
      d2y(1)=x-y(1)
      d2y(2)=x*x-y(2)
      END
```

Numerical Recipes provides two routines for stiff systems. The first, `stiff`, is a fourth-order Rosenbrock method. We try it out on the sample problem described in §16.6. Note that for best results one should change the line

```
      yscal(i)=abs(y(i))+abs(h*dydx(i))+TINY
```

in odeint to

```
      yscal(i)=max(1.,abs(y(i)))
```

```
      PROGRAM xstiff
C     driver for routine stiff
      INTEGER KMAXX,NMAX
      PARAMETER (KMAXX=200,NMAX=50)
      INTEGER kmax,kount,nbad,nok
      REAL dxsav,eps,hstart,x1,x2,y(3),xp,yp
      COMMON /path/ kmax,kount,dxsav,xp(KMAXX),yp(NMAX,KMAXX)
      EXTERNAL stiff,derivs
1     continue
      write(*,*) 'ENTER EPS,HSTART:'
      read(*,*,END=999) eps,hstart
      kmax=0
      x1=0.
      x2=50.
      y(1)=1.
      y(2)=1.
      y(3)=0.
      call odeint(y,3,x1,x2,eps,hstart,0.,nok,nbad,derivs,stiff)
      write(*,'(/1x,a,t30,i3)') 'Successful steps:',nok
      write(*,'(1x,a,t30,i3)') 'Bad steps:',nbad
      write(*,*) 'Y(END)=',y(1),y(2),y(3)
      goto 1
999   write(*,*) 'NORMAL COMPLETION'
      STOP
      END
```

The second method for stiff equations is a version of Bulirsch-Stoer integration that uses the semi-implicit midpoint rule `simpr` instead of `mmid`. Demonstration program `xsimpr` tries it out on the same test problem as above, exploring the effect of different stepsizes.

```
      PROGRAM xsimpr
C     driver for routine simpr
      INTEGER NMAX,NVAR
      REAL HTOT,X1
      PARAMETER(NMAX=3,NVAR=3,X1=0.,HTOT=50.)
      INTEGER i
      REAL a1,a2,a3,y(NMAX),yout(NMAX),
     *     dfdx(NMAX),dfdy(NMAX,NMAX),dydx(NMAX)
      EXTERNAL derivs
      y(1)=1.
      y(2)=1.
      y(3)=0.
      a1=0.5976
      a2=1.4023
      a3=0.
      call derivs(X1,y,dydx)
      call jacobn(X1,y,dfdx,dfdy,NVAR,NMAX)
      write(*,'(1x,a/)') 'Test Problem'
      do 11 i=5,50,5
        call simpr(y,dydx,dfdx,dfdy,NMAX,NVAR,X1,HTOT,i,yout,derivs)
        write(*,'(1x,a,f6.4,a,f7.4,a,i2,a)') 'X = ',X1,
     *       ' to ',X1+HTOT,' in ',i,' steps'
        write(*,'(1x,t5,a,t20,a)') 'Integration','Answer'
        write(*,'(1x,2f12.6)') yout(1),a1
        write(*,'(1x,2f12.6)') yout(2),a2
        write(*,'(1x,2f12.6)') yout(3),a3
11    continue
      END
```

Finally, sample program `xstifbs` solves the same problem with the full machinery of the Bulirsch-Stoer method via routine `stifbs`. Once again you should modify the scaling in `odeint` as in `xstiff` for best results.

```
      PROGRAM xstifbs
C     driver for routine stifbs
      INTEGER KMAXX,NMAX
      PARAMETER (KMAXX=200,NMAX=50)
      INTEGER kmax,kount,nbad,nok
      REAL dxsav,eps,hstart,x1,x2,y(3),xp,yp
      COMMON /path/ kmax,kount,dxsav,xp(KMAXX),yp(NMAX,KMAXX)
      EXTERNAL stifbs,derivs
1     continue
      write(*,*) 'ENTER EPS,HSTART:'
      read(*,*,END=999) eps,hstart
      kmax=0
      x1=0.
      x2=50.
      y(1)=1.
      y(2)=1.
      y(3)=0.
      call odeint(y,3,x1,x2,eps,hstart,0.,nok,nbad,derivs,stifbs)
      write(*,'(/1x,a,t30,i3)') 'Successful steps:',nok
      write(*,'(1x,a,t30,i3)') 'Bad steps:',nbad
```

```
      write(*,*) 'Y(END)=',y(1),y(2),y(3)
      goto 1
999   write(*,*) 'NORMAL COMPLETION'
      STOP
      END
```

Chapter 17: Two-Point Boundary Value Problems

Two-point boundary value problems, and their iterative solution, are the substance of Chapter 17 of *Numerical Recipes*. The first step in solving them is to cast the problem as a set of N coupled first-order ordinary differential equations, satisfying n_1 conditions at one boundary point, and $n_2 = N - n_1$ conditions at the other boundary point. We apply two general methods to the solutions. First are the shooting methods, typified by subroutines `shoot` and `shootf`, which enforce the n_1 conditions at one boundary and set n_2 conditions freely. Then they integrate across the interval to find discrepancies with the n_2 conditions to be enforced at the other end. The Newton-Raphson method is used to reduce these discrepancies by adjusting the variable parameters.

The other approach is the relaxation method in which the differential equations are replaced by finite difference equations on a grid that covers the range of interest. Routine `solvde` demonstrates this method, and is demonstrated "in action" by program `sfroid`, which uses it to compute eigenvalues of spheroidal harmonics. The programs `sphoot` and `sphfpt` in *Numerical Recipes* similarly demonstrate the use of `shoot` and `shootf` on the same problem. Since these three programs are essentially self-contained, we do not need separate demonstration programs here. The file `xsphfpt.f` on the diskette merely contains a copy of the subroutine `derivs` needed by `sphfpt`. We reproduce it here, along with typical output from the three test programs.

$$\star \quad \star \quad \star \quad \star$$

```
C       auxiliary routine for sphfpt
        SUBROUTINE derivs(x,y,dydx)
        INTEGER m,n
        REAL c2,dx,gamma,x,dydx(3),y(3)
        COMMON /sphcom/ c2,gamma,dx,m,n
        dydx(1)=y(2)
        dydx(2)=(2.0*x*(m+1.0)*y(2)-(y(3)-c2*x*x)*y(1))/(1.0-x*x)
        dydx(3)=0.0
        return
        END
```

In the following sample output, note that λ (`LAMBDA`) and μ (`mu`) are merely two different conventions for expressing the same result, related by

$$\mu(m,n) = \lambda(m,n) - m(m+1)$$

Typical output for `sfroid.f`:

```
ENTER M,N
3,5
ENTER C**2 OR 999 TO END
16.
      1      0.271210      1.000000
      2      0.080676      1.000000
      3      0.006255      1.000000
      4      0.000038      1.000000
      5      0.000000      1.000000
    M =       3  N =       5  C**2 =      16.00000      LAMBDA = 35.46101
ENTER C**2 OR 999 TO END
20.
      1      0.075821      1.000000
      2      0.005035      1.000000
      3      0.000024      1.000000
      4      0.000000      1.000000
    M =       3  N =       5  C**2 =      20.00000      LAMBDA = 36.77117
ENTER C**2 OR 999 TO END
-16.
      1      0.517707      1.000000
      2      0.835376      1.000000
      3      0.375956      1.000000
      4      0.172313      1.000000
      5      0.020226      1.000000
      6      0.000824      1.000000
      7      0.000001      1.000000
    M =       3  N =       5  C**2 =     -16.00000      LAMBDA = 24.31098
ENTER C**2 OR 999 TO END
-20.
      1      0.052213      1.000000
      2      0.003543      1.000000
      3      0.000022      1.000000
      4      0.000000      1.000000
    M =       3  N =       5  C**2 =     -20.00000      LAMBDA = 22.89307
ENTER C**2 OR 999 TO END
999
```

Typical output from sphoot.f:

```
input m,n,c-squared (999 to end)
3, 5, 1.5
    mu(m,n)
    18.524508
input m,n,c-squared (999 to end)
3, 5, 0.0
    mu(m,n)
    18.000002
input m,n,c-squared (999 to end)
3, 5, 999
```

Typical output from sphfpt.f:

```
input m,n,c-squared (999 to end)
3, 5, 1.5
    mu(m,n)
    18.524508
input m,n,c-squared (999 to end)
3, 5, -1.5
```

```
     mu(m,n)
    17.473288
 input m,n,c-squared (999 to end)
3, 5, 0.0
     mu(m,n)
    18.000004
 input m,n,c-squared (999 to end)
3, 5, 999
```

Chapter 18: Integral Equations and Inverse Theory

Linear Fredholm equations of the second kind can be solved with the routine `fred2`. *The accompanying routine* `fredin` *interpolates the solution returned by* `fred2` *on the quadrature points to arbitrary abscissas.* `voltra` *handles linear Volterra equations of the second kind. Routine* `wwghts` *constructs quadrature weights for an arbitrarily singular kernel. In Numerical Recipes a sample program demonstrating its use is given in* `fredex`, *so we do not give another sample program here.*

Fredholm equations of the first kind are closely related to so-called inverse problems. While no general-purpose routines are given in Numerical Recipes for such problems, there is an extensive discussion in Chapter 18.

$$\star \quad \star \quad \star \quad \star$$

Our test problem for routine `fred2` is Example 4.2.5b from *Computational Methods for Integral Equations* by Delves and Mohamed:

$$f(t) = \int_0^{\pi/2} (ts)^{3/4} f(s)\, ds + t^{1/2} - \left(\frac{\pi}{2}\right)^{9/4} \frac{t^{3/4}}{9/4}$$

with exact solution

$$f(t) = t^{1/2}$$

Using only an 8-point Gaussian quadrature gives reasonable accuracy in the solution.

```
      PROGRAM xfred2
C     driver for routine fred2
      INTEGER N
      REAL PI
      PARAMETER(N=8,PI=3.1415927)
      INTEGER i
      REAL a,ak,b,g,t(N),f(N),w(N)
      EXTERNAL g,ak
      a=0.
      b=PI/2.
      call fred2(N,a,b,t,f,w,g,ak)
C     compare with exact solution
      write(*,*) 'Abscissa, Calc soln, True soln'
      do 11 i=1,N
        write(*,100) t(i),f(i),sqrt(t(i))
11    continue
100   format(3f10.6)
      END

      REAL FUNCTION g(t)
```

```
      REAL PI
      PARAMETER(PI=3.1415927)
      REAL t
      g=sqrt(t)-(PI/2.)**2.25*t**.75/2.25
      return
      END


      REAL FUNCTION ak(t,s)
      REAL t,s
      ak=(t*s)**.75
      return
      END
```

Routine `fredin` interpolates the solution obtained with `fred2` onto an arbitrary point t using Nystrom interpolation. Sample program `xfredin` demonstrates its use on the same problem solved in `xfred2`.

```
      PROGRAM xfredin
C     driver for routine fredin
      INTEGER N
      REAL PI
      PARAMETER(N=8,PI=3.1415927)
      REAL a,ak,ans,b,fredin,g,x,t(N),f(N),w(N)
      EXTERNAL g,ak
      a=0.
      b=PI/2.
      call fred2(N,a,b,t,f,w,g,ak)
100   format(3f10.6)
1     write(*,*) 'Enter T between 0 and PI/2'
      read(*,*,END=99) x
      ans=fredin(x,N,a,b,t,f,w,g,ak)
      write(*,*) 'T, Calculated answer, True answer'
      write(*,100) x,ans,sqrt(x)
      go to 1
99    stop
      END


      REAL FUNCTION g(t)
      REAL PI
      PARAMETER(PI=3.1415927)
      REAL t
      g=sqrt(t)-(PI/2.)**2.25*t**.75/2.25
      return
      END


      REAL FUNCTION ak(t,s)
      REAL s,t
      ak=(t*s)**.75
      return
      END
```

Our test system of Volterra equations comes from p. 151 of Delves and Mohamed's book:

$$f_1(t) = -\int_0^t e^{t-s} f_1(s)\, ds - \int_0^t \cos(t-s) f_2(s)\, ds + \cosh t + t \sin t$$

$$f_2(t) = -\int_0^t e^{t+s} f_1(s)\, ds - \int_0^t t \cos s\, f_2(s)\, ds + 2\sin t + t(\sin^2 t + e^t)$$

with exact solution

$$f_1(t) = e^{-t}$$
$$f_2(t) = 2\sin t$$

```
      PROGRAM xvoltra
C     driver for routine voltra
      INTEGER N,M
      REAL H
      PARAMETER(N=30,H=.05,M=2)
      INTEGER nn
      REAL t0,t(N),f(M,N)
      EXTERNAL g,ak
      t0=0.
      call voltra(N,M,t0,H,t,f,g,ak)
C     exact soln is f(1)=exp(-t), f(2)=2sin(t)
      write(*,*)
     *'abscissa,voltra answer1,real answer1,voltra answer2,real answer2'
      do 11 nn=1,N
        write(*,100) t(nn),f(1,nn),exp(-t(nn)),f(2,nn),2.*sin(t(nn))
11    continue
100   format(5f10.6)
      END

      REAL FUNCTION g(k,t)
      INTEGER k
      REAL t
      if (k.eq.1) then
        g=cosh(t)+t*sin(t)
      else
        g=2.*sin(t)+t*(sin(t)**2+exp(t))
      endif
      return
      END

      REAL FUNCTION ak(k,l,t,s)
      INTEGER k,l
      REAL t,s
      if (k.eq.1) then
        if (l.eq.1) then
          ak=-exp(t-s)
        else
          ak=-cos(t-s)
        endif
      else
        if (l.eq.1) then
          ak=-exp(t+s)
        else
          ak=-t*cos(s)
        endif
      endif
```

```
return
END
```

Chapter 19: Partial Differential Equations

Several methods for solving partial differential equations by numerical means are treated in Chapter 19 of Numerical Recipes. All are finite differencing methods, including forward time centered space differencing, the Lax method, staggered leapfrog differencing, the two-step Lax-Wendroff scheme, the Crank-Nicholson method, Fourier analysis and cyclic reduction (FACR), Jacobi's method, the Gauss-Seidel method, successive over-relaxation (SOR) with and without Chebyshev acceleration, operator splitting methods as exemplified by the alternating direction implicit (ADI) method, and multigrid methods. There are so many methods, in fact, that we have not provided each topic with a subroutine of its own. In many cases the nature of such subroutines follows naturally from the description. In some cases, you will have to consult other references. The subroutines that do appear in the chapter, `sor`, `mglin` and `mgfas`, show two of the more useful and efficient methods for elliptic equations in application.

$$\star \quad \star \quad \star \quad \star$$

Subroutine `sor` incorporates successive over-relaxation with Chebyshev acceleration to solve an elliptic partial differential equation. As input it accepts six arrays of coefficients, an estimate of the spectral radius of Jacobi iteration, and a trial solution which is often just set to zero over the solution grid. In program `xsor` the method is applied to the model problem

$$\frac{\partial^2 u}{\partial x^2} + \frac{\partial^2 u}{\partial y^2} = \rho$$

which is treated as the relaxation problem

$$\frac{\partial u}{\partial t} = \frac{\partial^2 u}{\partial x^2} + \frac{\partial^2 u}{\partial y^2} - \rho$$

Using FTCS differencing, this becomes

$$u_{j+1,l}^n + u_{j-1,l}^n + u_{j,l+1}^n + u_{j,l-1}^n - 4u_{j,l}^{n+1} = \rho_{jl}\Delta^2$$

(The notation is explained in Chapter 19 of *Numerical Recipes*.) This is a simple form of the general difference equation to which `sor` may be applied, with

$$A_{jl} = B_{jl} = C_{jl} = D_{jl} = 1.0 \text{ and } E_{jl} = -4.0$$

for all j and l. The starting guess for u is $u_{jl} = 0.0$ for all j, l. We choose the density ρ to be zero everywhere except directly in the center of the grid where $\rho = 2$. For a 32×32 grid, $\Delta = 1/32$ and so $F_{j,l} = 0.0$ except in the center of the grid where

$F(\text{midl}, \text{midl}) = 2/32^2 \approx 0.0020$. The value of ρ_{Jacobi}, which is called `rjac`, is taken from equation (19.5.24) of *Numerical Recipes*,

$$\rho_{\text{Jacobi}} = \frac{\cos \dfrac{\pi}{J} + \left(\dfrac{\Delta x}{\Delta y}\right)^2 \cos \dfrac{\pi}{L}}{1 + \left(\dfrac{\Delta x}{\Delta y}\right)^2}$$

In this case, j=l=JMAX and $\Delta x = \Delta y$ so rjac = $\cos(\pi/\text{JMAX})$. A call to `sor` leads to the solution shown below. To fit the output on the page, the solution at only every fourth grid point is printed out. As a test that this is indeed a solution to the finite difference equation, the program plugs the result back into that equation, calculating

$$F_{j,l} = u^n_{j+1,l} + u^n_{j-1,l} + u^n_{j,l+1} + u^n_{j,l-1} - 4u^{n+1}_{j,l}$$

The test is whether $F_{j,l}$ is almost everywhere zero, but equal to 0.0020 at the very centerpoint of the grid.

```
SOR Solution:
 0.0000  0.0000  0.0000  0.0000  0.0000  0.0000  0.0000  0.0000
 0.0000  0.0000 -0.0001 -0.0001 -0.0001 -0.0001 -0.0001  0.0000  0.0000
 0.0000 -0.0001 -0.0001 -0.0002 -0.0002 -0.0002 -0.0001 -0.0001  0.0000
 0.0000 -0.0001 -0.0002 -0.0003 -0.0005 -0.0003 -0.0002 -0.0001  0.0000
 0.0000 -0.0001 -0.0002 -0.0005 -0.0014 -0.0005 -0.0002 -0.0001  0.0000
 0.0000 -0.0001 -0.0002 -0.0003 -0.0005 -0.0003 -0.0002 -0.0001  0.0000
 0.0000 -0.0001 -0.0001 -0.0002 -0.0002 -0.0002 -0.0001 -0.0001  0.0000
 0.0000  0.0000 -0.0001 -0.0001 -0.0001 -0.0001 -0.0001  0.0000  0.0000
 0.0000  0.0000  0.0000  0.0000  0.0000  0.0000  0.0000  0.0000  0.0000
```

```
      PROGRAM xsor
C     driver for routine sor
      INTEGER JMAX,NSTEP
      REAL PI
      PARAMETER(JMAX=33,NSTEP=4,PI=3.1415926)
      INTEGER i,j,midl
      DOUBLE PRECISION rjac,a(JMAX,JMAX),b(JMAX,JMAX),c(JMAX,JMAX),
     *     d(JMAX,JMAX),e(JMAX,JMAX),f(JMAX,JMAX),u(JMAX,JMAX)
      do 12 i=1,JMAX
        do 11 j=1,JMAX
          a(i,j)=1.d0
          b(i,j)=1.d0
          c(i,j)=1.d0
          d(i,j)=1.d0
          e(i,j)=-4.d0
          f(i,j)=0.d0
          u(i,j)=0.d0
11      continue
12    continue
      midl=JMAX/2+1
      f(midl,midl)=2.d0/(JMAX-1)**2
      rjac=cos(PI/JMAX)
      call sor(a,b,c,d,e,f,u,JMAX,rjac)
      write(*,'(1x,a)') 'SOR Solution:'
      do 13 i=1,JMAX,NSTEP
        write(*,'(1x,9f8.4)') (u(i,j),j=1,JMAX,NSTEP)
```

```
13     continue
       write(*,'(/1x,a)') 'Test that solution satisfies Difference Eqns:'
       do 15 i=NSTEP+1,JMAX-1,NSTEP
         do 14 j=NSTEP+1,JMAX-1,NSTEP
           f(i,j)=u(i+1,j)+u(i-1,j)+u(i,j+1)+u(i,j-1)-4.d0*u(i,j)
14       continue
         write(*,'(7x,9f8.4)') (f(i,j),j=NSTEP+1,JMAX-1,NSTEP)
15     continue
       END
```

Multigrid methods are generally much more efficient than SOR. Routine mglin solves linear elliptic equations using the full multigrid method. Program xmglin solves the same model problem as xsor and prints out the solution, which may be compared with the copy listed above. However, because of the way mglin is set up, when the solution is substituted into the difference equations, it should give back ρ and not $\rho\Delta^2$. Here is the output on a 32×32 grid:

```
Test that solution satisfies Difference Eqns:
     0.0001 -0.0001 -0.0004  0.0008 -0.0004 -0.0001  0.0001
    -0.0001 -0.0002 -0.0008  0.0028 -0.0008 -0.0002 -0.0001
    -0.0004 -0.0008 -0.0027  0.0048 -0.0027 -0.0008 -0.0004
     0.0008  0.0028  0.0048  2.0171  0.0048  0.0028  0.0008
    -0.0004 -0.0008 -0.0027  0.0048 -0.0027 -0.0008 -0.0004
    -0.0001 -0.0002 -0.0008  0.0028 -0.0008 -0.0002 -0.0001
     0.0001 -0.0001 -0.0004  0.0008 -0.0004 -0.0001  0.0001
```

```
       PROGRAM xmglin
C      driver for routine mglin
       INTEGER JMAX,NSTEP
       PARAMETER(JMAX=33,NSTEP=4)
       INTEGER i,j,midl
       DOUBLE PRECISION f(JMAX,JMAX),u(JMAX,JMAX)
       do 12 i=1,JMAX
         do 11 j=1,JMAX
           u(i,j)=0.d0
11       continue
12     continue
       midl=JMAX/2+1
       u(midl,midl)=2.d0
       call mglin(u,JMAX,2)
       write(*,'(1x,a)') 'MGLIN Solution:'
       do 13 i=1,JMAX,NSTEP
         write(*,'(1x,9f8.4)') (u(i,j),j=1,JMAX,NSTEP)
13     continue
       write(*,'(/1x,a)') 'Test that solution satisfies Difference Eqns:'
       do 15 i=NSTEP+1,JMAX-1,NSTEP
         do 14 j=NSTEP+1,JMAX-1,NSTEP
           f(i,j)=u(i+1,j)+u(i-1,j)+u(i,j+1)+u(i,j-1)-4.d0*u(i,j)
14       continue
         write(*,'(7x,7f8.4)')
     *         (f(i,j)*(JMAX-1)*(JMAX-1),j=NSTEP+1,JMAX-1,NSTEP)
15     continue
       END
```

Numerical Recipes provides mgfas as a prototype for solving nonlinear elliptic problems with the Full Approximation Storage (FAS) scheme. Program xmgfas solves the simple problem

$$\nabla^2 u + u^2 = \rho$$

as described in *Numerical Recipes* with the same choice of ρ made in xsor and xmglin. The solution is checked once again by substituting into the difference equation. Here is the output for the 32×32 grid:

```
MGFAS Solution:
  0.0000  0.0000  0.0000  0.0000  0.0000  0.0000  0.0000  0.0000  0.0000
  0.0000  0.0000 -0.0001 -0.0001 -0.0001 -0.0001 -0.0001  0.0000  0.0000
  0.0000 -0.0001 -0.0001 -0.0002 -0.0002 -0.0002 -0.0001 -0.0001  0.0000
  0.0000 -0.0001 -0.0002 -0.0004 -0.0004 -0.0004 -0.0002 -0.0001  0.0000
  0.0000 -0.0001 -0.0002 -0.0004 -0.0013 -0.0004 -0.0002 -0.0001  0.0000
  0.0000 -0.0001 -0.0002 -0.0004 -0.0004 -0.0004 -0.0002 -0.0001  0.0000
  0.0000 -0.0001 -0.0001 -0.0002 -0.0002 -0.0002 -0.0001 -0.0001  0.0000
  0.0000  0.0000 -0.0001 -0.0001 -0.0001 -0.0001 -0.0001  0.0000  0.0000
  0.0000  0.0000  0.0000  0.0000  0.0000  0.0000  0.0000  0.0000  0.0000
Test that solution satisfies Difference Eqns:
      0.0000  0.0008 -0.0009  0.0000 -0.0009  0.0008  0.0000
      0.0008  0.0033  0.0007 -0.0035  0.0007  0.0033  0.0008
     -0.0009  0.0007  0.0051 -0.0328  0.0051  0.0007 -0.0009
      0.0000 -0.0035 -0.0328  1.8660 -0.0328 -0.0035  0.0000
     -0.0009  0.0007  0.0051 -0.0328  0.0051  0.0007 -0.0009
      0.0008  0.0033  0.0007 -0.0035  0.0007  0.0033  0.0008
      0.0000  0.0008 -0.0009  0.0000 -0.0009  0.0008  0.0000
```

You might be a little puzzled by this output at first. For example, the center value of $\nabla^2 u + u^2$ for the solution is 1.8660, while the true value is 2. Both sor and mglin get closer, with 2.0171, yet the nonlinear correction u^2 is only $(-.0013)^2$. Here's what's going on: mgfas iterates until the iteration error is smaller than the truncation error. These errors are r.m.s. errors, computed with routine anorm2. To check the performance of mgfas, we did a run with the parameter ALPHA set to 10^{-6} and with maxcyc $= 6$. This high accuracy solution is essentially the exact solution of the finite difference equations, and is in fact very close to the solution obtained by sor and mglin. We next computed the truncation error by applying the routine lop to this accurate solution and then finding the difference between the answer and the original right-hand side, ρ. The norm of this error was 6×10^{-2}. Next, we applied lop to the solution obtained in the demonstration program. (This gives the quantity printed above as "Test that solution satisfies Difference Eqns:".) The difference between this quantity and lop of the "exact" finite difference solution is the iteration error. Its norm is 6×10^{-3}. Thus mgfas is solving the problem to better than the truncation error, as advertised. Since in this problem ρ is so sharply peaked, the r.m.s. error is perhaps not the best measure of the error — the maximum deviation might be better. For a smoother ρ, these error measures are not that different and mgfas would give more intuitive looking results. For this problem, however, you need a smaller value of ALPHA if you want the maximum deviation below the truncation error; ALPHA $= 0.1$ will do the trick.

The routine maloc is reproduced here in the form required for use with mgfas. The main book reproduces the slightly different form that is used with mglin.

```
      PROGRAM xmgfas
C     driver for routine mgfas
      INTEGER JMAX,NSTEP
      PARAMETER(JMAX=33,NSTEP=4)
```

```
      INTEGER i,j,midl
      DOUBLE PRECISION f(JMAX,JMAX),u(JMAX,JMAX)
      do 12 i=1,JMAX
        do 11 j=1,JMAX
        u(i,j)=0.d0
11      continue
12    continue
      midl=JMAX/2+1
      u(midl,midl)=2.d0
      call mgfas(u,JMAX,2)
      write(*,'(1x,a)') 'MGFAS Solution:'
      do 13 i=1,JMAX,NSTEP
        write(*,'(1x,9f8.4)') (u(i,j),j=1,JMAX,NSTEP)
13    continue
      write(*,'(/1x,a)') 'Test that solution satisfies Difference Eqns:'
      do 15 i=NSTEP+1,JMAX-1,NSTEP
        do 14 j=NSTEP+1,JMAX-1,NSTEP
        f(i,j)=u(i+1,j)+u(i-1,j)+u(i,j+1)+u(i,j-1)-4.d0*u(i,j)
     *        +u(i,j)**2/(JMAX-1)**2
14      continue
        write(*,'(7x,7f8.4)')
     *      (f(i,j)*(JMAX-1)*(JMAX-1),j=NSTEP+1,JMAX-1,NSTEP)
15    continue
      END


      FUNCTION maloc(len)
      INTEGER maloc,len,NG,MEMLEN
      PARAMETER (NG=5,MEMLEN=17*2**(2*NG)/3+18*2**NG+10*NG-86/3)
      INTEGER mem
      DOUBLE PRECISION z
      COMMON /memory/ z(MEMLEN),mem
      if (mem+len+1.gt.MEMLEN) pause 'insufficient memory in maloc'
      z(mem+1)=len
      maloc=mem+2
      mem=mem+len+1
      return
      END
```

Chapter 20: Less-Numerical Algorithms

Chapter 20 of Numerical Recipes collects together an idiosyncratic group of "less-numerical" algorithms whose usefulness lies either in tasks peripheral to numerical work, or in elucidating some less obvious applications of numerical techniques (for example, the use of Fast Fourier Transforms for performing extended precision arithmetic).

The routine machar is a useful aid in understanding the limits of floating point arithmetic on your particular computer. igray is a simple function that converts integers into their Gray code, or back again. The combination of icrc and its captive routine icrc1 is used to construct cyclic redundancy checksums, in some standard conventions. The main book discusses two rather different methods for compressing information, Huffman coding, realized in the routines hufmak, hufapp, hufenc, and hufdec, and arithmetic coding, realized as arcmak, arcode, and arcsum. Finally, a whole group of routines for multiple precision arithmetic are given, with names of the form mp..., culminating in the program mppi which is used to compute π to as many as several thousand decimal digits.

$$\star \quad \star \quad \star \quad \star$$

The sample program for machar does nothing more than run the routine and print the results. machar, by some fairly subtle arithmetic processes, feels out the floating point characteristics of your computer. The main book explains the meaning of the parameters that are reported. The most important ones are eps, xmin, and xmax, which give respectively the floating point precision, and the smallest and largest (in magnitude) numbers that are representable at full precision.

```
      PROGRAM xmachar
C     driver for routine machar
      INTEGER ibeta,iexp,irnd,it,machep,maxexp,minexp,negep,ngrd
      REAL eps,epsneg,xmax,xmin
      call machar(ibeta,it,irnd,ngrd,machep,negep,iexp,minexp,
     *     maxexp,eps,epsneg,xmin,xmax)
      write(*,*) 'ibeta=',ibeta
      write(*,*) 'it=',it
      write(*,*) 'irnd=',irnd
      write(*,*) 'ngrd=',ngrd
      write(*,*) 'machep=',machep
      write(*,*) 'negep=',negep
      write(*,*) 'iexp=',iexp
      write(*,*) 'minexp=',minexp
      write(*,*) 'maxexp=',maxexp
      write(*,*) 'eps=',eps
```

```
      write(*,*) 'epsneg=',epsneg
      write(*,*) 'xmin=',xmin
      write(*,*) 'xmax=',xmax
      END
```

The function `igray` returns the Gray code of its first argument, or else the inverse Gray code (if the second argument is -1). The following example program verifies, for some range of integers, that the function, followed by its inverse, does in fact produce the original input value. It also prints out a few values within the chosen range, and verifies, for those values, that the Gray code of two consecutive integers differ in only one bit: the last column printed should always be a power of 2; half the time it will simply be the value 1, since the results differ only in the least significant bit.

```
      PROGRAM xigray
C     driver for routine igray
      INTEGER igray,jp,n,ng,nmax,nmin,nni,nxor
1     write(*,*) 'INPUT NMIN,NMAX:'
      read(*,*,END=999) nmin,nmax
      jp=max(1,(nmax-nmin)/11)
      write(*,*) 'n, Gray(n), Gray(Gray(n)), Gray(n).xor.Gray(n+1)'
      do 11 n=nmin,nmax
        ng=igray(n,1)
        nni=igray(ng,-1)
        if (nni.ne.n) write(*,*) 'WRONG ! AT ',n,ng,nni
        if (mod(n-nmin,jp).eq.0) then
          nxor=ieor(ng,igray(n+1,1))
          write(*,*) n,ng,nni,nxor
        endif
11    continue
      goto 1
999   write(*,*) 'NORMAL COMPLETION'
      STOP
      END
```

Cyclic redundancy checksums are constructed, according to any one of several standard conventions, by the routine `icrc`. The following test program, with the input lines

```
001T
017CatMouse987654321
```

will reproduce the test values that are given in the Table in the main book. Notice how some conventions (e.g., X.25), require that the packet (message and checksum together) be made from the bitwise complement of the CRC bytes, while others use the bytes themselves. Note also that the desired checksum of the full packet is zero in some conventions, but some particular value (here hexadecimal F0B8) in others. The sample program uses the utility `hexout`, discussed earlier with the sample program for `psdes` in Chapter 7.

```
      PROGRAM xicrc
C     driver for routine icrc
      INTEGER i1,i2,j,n
      INTEGER icrc,iand
      CHARACTER*10 hexout
```

```
      CHARACTER*1 lin(80),chbyt(2)
1     write(*,*) 'ENTER LENGTH,STRING: '
      read(*,'(i3,80a1)',END=999) n,(lin(j),j=1,n)
      write(*,*) (lin(j),j=1,n)
      i1=icrc(chbyt,lin,n,0,1)
      lin(n+1)=chbyt(2)
      lin(n+2)=chbyt(1)
      i2=icrc(chbyt,lin,n+2,0,1)
      write(*,'('' XMODEM: String CRC, Packet CRC='',2a12)')
*         hexout(i1),hexout(i2)
      i1=icrc(chbyt,lin,n,255,-1)
      lin(n+1)=char(iand(255,not(ichar(chbyt(1)))))
      lin(n+2)=char(iand(255,not(ichar(chbyt(2)))))
      i2=icrc(chbyt,lin,n+2,255,-1)
      write(*,'(''   X.25: String CRC, Packet CRC='',2a12)')
*         hexout(i1),hexout(i2)
      i1=icrc(chbyt,lin,n,0,-1)
      lin(n+1)=chbyt(1)
      lin(n+2)=chbyt(2)
      i2=icrc(chbyt,lin,n+2,0,-1)
      write(*,'('' CRC-CCITT: String CRC, Packet CRC='',2a12)')
*         hexout(i1),hexout(i2)
      goto 1
999   write(*,*) 'NORMAL COMPLETION'
      STOP
      END


      FUNCTION hexout(num)
C     Numerical Recipes Fortran utility routine for printing out hexadecimal
      CHARACTER*10 hexout
      INTEGER NCOMP
      PARAMETER (NCOMP=268435455)
      INTEGER i,j,n,num
      CHARACTER*1 hexit(16)
      SAVE hexit
      DATA hexit /'0','1','2','3','4','5','6','7','8','9',
*        'a','b','c','d','e','f'/
      n=num
      if (n.lt.0) then
        i=mod(n,16)
        if (i.lt.0) i=16+i
        n=n/16
        n=-n
        if (i.ne.0) then
          n=NCOMP-n
        else
          n=NCOMP+1-n
        endif
      else
        i=mod(n,16)
        n=n/16
      endif
      j=10
      hexout(j:j)=hexit(i+1)
2     if (n.gt.0) then
        i=mod(n,16)
        n=n/16
```

```
      j=j-1
      hexout(j:j)=hexit(i+1)
      goto 2
      endif
      j=j-1
      hexout(j:j)='x'
      j=j-1
      hexout(j:j)='0'
      do 11 i=j-1,1,-1
        hexout(i:i)=' '
11    continue
      return
      END
```

The example program for the decimal check digit routine decchk first verifies the claim in the main book that the method finds about 95% of jump transpositions ($acb \to bca$). (The exact answer is that it finds all but 52 out of 900.) The program then prompts the user for test strings (terminated by "x") and prints out the string with a following check digit, and verification that the augmented string passes the check-digit test. You might test this program with, e.g., telephone numbers in various formats, such as "1-800-433-7300" or "1 (800) 241-6522," to verify that decchk in fact ignores all input characters except decimal digits in computing its check digit.

```
      PROGRAM xdecchk
C     driver for routine decchk
      INTEGER j,k,l,n,nbad,ntot
      LOGICAL decchk,iok,jok
      CHARACTER lin*128,ch*1,chh*1,tf*1
C     test all jump transpositions of the form 86jlk41
      ntot=0
      nbad=0
      do 13 j=48,57
        do 12 k=48,57
          do 11 l=48,57
            if (j.ne.k) then
              ntot=ntot+1
              lin(1:7)='86'//char(j)//char(l)//char(k)//'41'
              iok=decchk(lin(1:7),7,ch)
              iok=decchk(lin(1:7)//ch,8,chh)
              lin(1:7)='86'//char(k)//char(l)//char(j)//'41'
              jok=decchk(lin(1:7)//ch,8,chh)
              if ((.not.iok).or.jok) then
                nbad=nbad+1
              endif
            endif
11        continue
12      continue
13    continue
      write(*,'(1x,a,t30,i3)') 'Total tries:',ntot
      write(*,'(1x,a,t30,i3)') 'Bad tries:',nbad
      write(*,'(1x,a,t30,f3.2)') 'Fraction good:',
     *   float(ntot-nbad)/float(ntot)
C     construct check digits for some user-supplied strings
1     write(*,*) 'enter string terminated by x:'
      read(*,'(a20)',END=999) lin
      do 14 j=1,128
```

```
        if (lin(j:j).eq.'x') goto 2
14      continue
2       n=j-1
        if (n.eq.0) goto 999
        iok=decchk(lin(1:n),n,ch)
        jok=decchk(lin(1:n)//ch,n+1,chh)
        if (jok) then
          tf='T'
        else
          tf='F'
        endif
        write(*,*) lin(1:n)//ch,' checks as ',tf
        goto 1
999     write(*,*) 'NORMAL COMPLETION'
        STOP
        END
```

The example program that demonstrates Huffman compression of alphabetic text has several separate tasks to perform. First, it must construct a statistical profile of the kind of text to be compressed, in the form of a table of letter frequencies. It does this using the file TEXT.DAT which, as supplied on the diskette, contains a particularly racy passage from Thomas Hardy's *Far from the Madding Crowd* (New York: Harper, 1895).

Next, the Huffman code itself is constructed from the frequency count. The demonstration program prints out the entire code, allowing you to see how more common letters are encoded in fewer bits than less common ones.

The program then prompts repeatedly for an input line of text, compresses that line, decompresses it, and prints out the result (hopefully the same as you input) and the lengths of the line in uncompressed and compressed forms. Notice how a compressed line is terminated by encoding a (spurious) rare character, while *not* incrementing the message byte count. This ensures that on decoding, the spurious character will not be fully decoded. In addition to trying ordinary text as input to the program, you might try a line of all lower case e's, or all upper case X's.

The demonstration program is unfortunately cluttered by workarounds to one of ANSI FORTRAN's classic bugaboos, its inability to tell how many characters are in an input line of text. We must loop backwards from the end of the input buffer assuming that the input line has no trailing blanks.

```
        PROGRAM xhuffman
C       driver for routines hufmak, hufenc, hufdec
        INTEGER MC,MQ,MAXBUF,MAXLINE
        PARAMETER (MC=512,MQ=2*MC-1,MAXBUF=200,MAXLINE=80)
        INTEGER icod,ncod,nprob,left,iright,nch,nodemax
        COMMON /hufcom/ icod(MQ),ncod(MQ),nprob(MQ),left(MQ),
     *      iright(MQ),nch,nodemax
        SAVE /hufcom/
        INTEGER i,ilong,j,k,n,nb,nh,nlong,nt,nfreq(256)
        CHARACTER*200 mess,code,ness
        CHARACTER*80 lin
C       construct a letter frequency table from the file TEXT.DAT
        open(unit=7,file='TEXT.DAT',status='old')
        do 11 j=1,256
          nfreq(j)=0
```

```
11      continue
1       continue
        do 12 j=1,MAXLINE
          lin(j:j)=char(32)
12      continue
        read(7,'(a)',END=3) lin
        do 13 n=MAXLINE,1,-1
          if (lin(n:n).ne.char(32)) goto 2
13      continue
2       do 14 k=1,min(MAXLINE,n)
          j=ichar(lin(k:k))-31
          if (j.ge.1) nfreq(j)=nfreq(j)+1
14      continue
        goto 1
3       close(unit=7)
        nch=96
C       here is the initialization that constructs the code
        call hufmak(nfreq,nch,ilong,nlong)
        write(*,*)
     *      'index, ','character, ','nfreq, ','bits in code, ','code int'
        do 15 j=1,nch
          if (nfreq(j).ne.0.)
     *        write(*,*) j,' ',char(j+31),' ',nfreq(j),ncod(j),icod(j)
15      continue
C       now ready to prompt for lines to encode
4       write(*,*) 'ENTER A LINE:'
        do 16 j=1,MAXLINE
          mess(j:j)=char(32)
16      continue
        read(*,'(a)',END=999) mess
        do 17 n=MAXLINE,1,-1
          if (mess(n:n).ne.char(32)) goto 5
17      continue
C       shift from 256 character alphabet to 96 printing characters
5       do 18 j=1,n
          mess(j:j)=char(ichar(mess(j:j))-32)
18      continue
C       here we Huffman encode mess(1:n)
        nb=0
        do 19 j=1,n
          call hufenc(ichar(mess(j:j)),code,MAXLINE,nb)
19      continue
        nh=nb/8+1
C       message termination (encode a single long character)
        call hufenc(ilong,code,MAXLINE,nb)
C       here we decode the message, hopefully to get the original back
        nb=0
        do 21 j=1,MAXLINE
          call hufdec(i,code,nh,nb)
          if (i.eq.nch) goto 6
          ness(j:j)=char(i)
21      continue
        pause 'HUFFMAN - NEVER GET HERE'
6       nt=j-1
        write(*,*) 'LENGTH OF LINE INPUT,CODED=',n,nh
        write(*,*) 'DECODED OUTPUT:'
        write(*,'(1x,80a1)') (char(ichar(ness(j:j))+32),j=1,nt)
```

```
      if (nt.ne.n) write(*,*) 'ERROR ! N DECODED .NE. N INPUT'
      if (nt-n.eq.1) write(*,*) 'MAY BE HARMLESS SPURIOUS CHARACTER.'
      goto 4
999   write(*,*) 'NORMAL COMPLETION'
      STOP
      END
```

The demonstration program for arithmetic compression is very similar to that for Huffman compression. As before, we construct a letter frequency table from a corpus that is statistically similar to the text that we are expecting to compress, then initialize the code, and finally encode/decode messages.

There are several characters' worth of output associated with message initialization and message termination in arithmetic coding, so one would not normally begin a new compression for each input line (as is done in the following demonstration). Rather, a whole file would normally be compressed in a single pass.

```
      PROGRAM xarcode
C     driver for routines arcmak, arcode
      INTEGER MC,MD,MQ,NWK,MAXBUF,MAXLINE
      PARAMETER (MC=512,MD=MC-1,MQ=2*MC-1,NWK=20,MAXBUF=200,MAXLINE=80)
      INTEGER nch,ncum,nrad,minint,jdif,nc,ilob,iupb,ncumfq
      INTEGER i,j,k,lc,n,nt,nfreq(256)
      COMMON /arccom/ ncumfq(MC+2),iupb(NWK),ilob(NWK),nch,nrad,
     *     minint,jdif,nc,ncum
      SAVE /arccom/
      CHARACTER*1 code(MAXBUF)
      CHARACTER*80 lin
      CHARACTER*200 mess,ness
      open(unit=7,file='TEXT.DAT',status='old')
      do 11 j=1,256
        nfreq(j)=0
11    continue
1     continue
      do 12 j=1,MAXLINE
        lin(j:j)=char(32)
12    continue
      read(7,'(a)',END=3) lin
      do 13 n=MAXLINE,1,-1
        if (lin(n:n).ne.char(32)) goto 2
13    continue
2     do 14 k=1,min(MAXLINE,n)
        j=ichar(lin(k:k))-31
        if (j.ge.1) nfreq(j)=nfreq(j)+1
14    continue
      goto 1
3     close(unit=7)
      nch=96
      nrad=256
C     here is the initialization that constructs the code
      call arcmak(nfreq,nch,nrad)
C     now ready to prompt for lines to encode
4     write(*,*) 'ENTER A LINE:'
      do 15 j=1,MAXLINE
        mess(j:j)=char(32)
15    continue
      read(*,'(a)',END=999) mess
```

```
        do 16 n=MAXLINE,1,-1
          if (mess(n:n).ne.char(32)) goto 5
16      continue
C       shift from 256 character alphabet to 96 printing characters
5       do 17 j=1,n
          mess(j:j)=char(ichar(mess(j:j))-32)
17      continue
C       message initialization
        lc=1
        call arcode(0,code,MAXBUF,lc,0)
C       here we arithmetically encode mess(1:n)
        do 18 j=1,n
          call arcode(ichar(mess(j:j)),code,MAXBUF,lc,1)
18      continue
        call arcode(nch,code,MAXBUF,lc,1)
C       message termination
        write(*,*) 'LENGTH OF LINE INPUT, CODED=',n,lc-1
C       here we decode the message, hopefully to get the original back
        lc=1
        call arcode(0,code,MAXBUF,lc,0)
        do 19 j=1,MAXBUF
          call arcode(i,code,MAXBUF,lc,-1)
          if (i.eq.nch) goto 6
          ness(j:j)=char(i)
19      continue
        pause 'ARCODE - NEVER GET HERE'
6       nt=j-1
        write(*,*) 'DECODED OUTPUT:'
        write(*,'(1x,80a1)') (char(ichar(ness(j:j))+32),j=1,nt)
        if (nt.ne.n) write(*,*) 'ERROR ! J DECODED .NE. N INPUT',j,n
        goto 4
999     write(*,*) 'NORMAL COMPLETION'
        STOP
        END
```

The *Numerical Recipes* book ends on a note of numerical denouement with the calculation of π to some 2398 decimal places. The following demonstration program, and its one additional subroutine, repeats that calculation, and also uses the various multiple precision routines to calculate a couple of simpler quantities at high precision, $\sqrt{2}$ and $2 - \sqrt{2}$. You might wonder why bother with the latter quantity. The answer is that the multiple precision integer divide routine mpdiv happens not to be used at all in the calculation of π (although the routine for reciprocals, mpinv, is used). To give it some exercise, we therefore calculate $2 - \sqrt{2}$ by the following nutty scheme that does make use of the integer divide:

$$\frac{2 \times 2^M}{[\sqrt{2} \times 2^M]} = 1 \text{ remainder } (2 - \sqrt{2}) \times 2^M$$

at least to the same approximation that $\sqrt{2}$ is computed. Here square brackets indicate integer part, and M is the number of binary digits in the precision that you supply when prompted.

Before you go off and compute a million digits of π, be sure to read the main book's note about the routine mp2dfr for converting from base 256 to decimal fractions. Unless you are satisfied with knowing π in base 256 (or binary, octal, or hexadecimal), you will need to rewrite that routine to make it faster.

```
      PROGRAM xmppi
C     driver for mp routines
      INTEGER n
3     write(*,*) 'INPUT N'
      read(*,*,END=999) n
      call mpinit
      call mpsqr2(n)
      call mppi(n)
      goto 3
999   write(*,*) 'NORMAL COMPLETION'
      STOP
      END

      SUBROUTINE mpsqr2(n)
      INTEGER IAOFF,NMAX
      PARAMETER (IAOFF=48,NMAX=8192)
      INTEGER j,n,m
      CHARACTER*1 x(NMAX),y(NMAX),t(NMAX),s(3*NMAX),q(NMAX),r(NMAX)
      t(1)=char(2)
      do 11 j=2,n
        t(j)=char(0)
11    continue
      call mpsqrt(x,x,t,n,n)
      call mpmov(y,x,n)
      write(*,*) 'SQRT(2)='
      s(1)=char(ichar(y(1))+IAOFF)
      s(2)='.'
C     caution: next step is N**2! omit it for large N
      call mp2dfr(y(2),s(3),n-1,m)
      write(*,'(1x,64a1)') (s(j),j=1,m+2)
      write(*,*) 'Result rounded to 1 less base-256 place:'
C     use s as scratch space
      call mpsad(s,x,n,128)
      call mpmov(y,s(2),n-1)
      s(1)=char(ichar(y(1))+IAOFF)
      s(2)='.'
C     caution: next step is N**2! omit it for large N
      call mp2dfr(y(2),s(3),n-2,m)
      write(*,'(1x,64a1)') (s(j),j=1,m+2)
      write(*,*) '2-SQRT(2)='
C     calculate this the hard way to exercise the mpdiv function
      call mpdiv(q,r,t,x,n,n)
      s(1)=char(ichar(r(1))+IAOFF)
      s(2)='.'
C     caution: next step is N**2! omit it for large N
      call mp2dfr(r(2),s(3),n-1,m)
      write(*,'(1x,64a1)') (s(j),j=1,m+2)
      return
      END
```

Index of Demonstrated Subroutines

Following is a brief explanation of each *Numerical Recipes* product, plus two order forms (one for North American residents, one for all other), which may be used to order these items directly from the publisher if you cannot obtain them from your local bookstore.

Numerical Recipes in FORTRAN, Second Edition and *Numerical Recipes in C, Second Edition* represent the main text and reference component of the *Numerical Recipes* package. Each book contains over 300 programs, in the language of the reader's choice, and constitutes a complete subroutine library for scientific computation. Both versions contain equivalent tutorial, mathematical, and practical discussions.

There are two example books containing FORTRAN or C source programs respectively that exercise and demonstrate all of the *Numerical Recipes, Second Edition* programs. Each example program contains comments and is prefaced by a short description. Input and output data are supplied in many cases. The example books are designed to help readers incorporate procedures and subroutines and conduct simple validation tests.

The programs contained in both the second edition main books and the example books are available in several machine-readable formats that will save users hours of tedious keyboarding. Diskettes for IBM PC compatible machines are available in either 5¼ inch high density or 3½ inch format and operate on DOS 2.0 or later. Diskettes for the Apple Macintosh are 3½ inch single-sided disks.

Some selected first edition products are also still available:

Numerical Recipes in Pascal contains the original 200 *Numerical Recipes* routines translated into Pascal along with the tutorial text. The Pascal example book contains the example programs for these routines. Diskettes for each book are available in 5¼ inch double-sided, double-density IBM and 3½ inch Apple Macintosh format.

Numerical Recipes Routines and Examples in BASIC contains all the routines from the original Numerical Recipes plus the exercise programs from the example book, all translated into BASIC, along with the text from the example book. The BASIC routines and programs are also available on 5 ¼ inch diskette for IBM PC compatible machines.

Instructions

To obtain the books or the latest version of the disks, please order from your bookstore or complete the information on the order form in this book and mail it to Cambridge University Press in Port Chester, New York or Cambridge, England. Please note that there is a separate order form for each location. All orders must be prepaid. Ordinary postage for shipping is paid by the publisher.

NB: Technical questions, corrections, and requests for information on mainframe and workstation licenses should be directed to Numerical Recipes Software, P.O. Box 243, Cambridge, MA 02238, U.S.A. Please do not write the publisher.

There are no cash refunds for diskettes. Only diskettes with manufacturing defects may be returned to the publisher for replacement.

ORDER FORM (United States and Canada)

Order from your bookstore or mail to
Cambridge University Press, Order Department, 110 Midland Avenue, Port Chester,
New York 10573
Call for current prices: 1-800-431-1580

.... 43064-X   Numerical Recipes in FORTRAN: The Art of Scientific Computing,
                     Second Edition
.... 43717-2   FORTRAN Diskette (IBM, 5¼ inch/1.2M), second edition
.... 43719-9   FORTRAN Diskette (IBM, 3½ inch/720K), second edition
.... 43716-4   FORTRAN Diskette (Macintosh), second edition
.... 43721-0   FORTRAN Example Book, second edition

.... 43108-5   Numerical Recipes in C: The Art of Scientific Computing, Second Edition
.... 43714-8   C Diskette (IBM, 5¼ inch/1.2M), second edition
.... 43724-5   C Diskette (IBM, 3½ inch/720K), second edition
.... 43715-6   C Diskette (Macintosh), second edition
.... 43720-2   C Example Book, second edition

The following first edition products are still available:

.... 37516-9   Numerical Recipes in Pascal: The Art of Scientific Computing
.... 37532-0   Pascal Diskette (IBM)
.... 38766-3   Pascal Diskette (Macintosh)
.... 37675-0   Pascal Example Book
.... 37533-9   Pascal Example Diskette (IBM)
.... 38767-1   Pascal Example Diskette (Macintosh)

.... 40689-7   Numerical Recipes Routines and Examples in BASIC
.... 40688-9   BASIC Diskette (IBM)

Please indicate method of payment: check ...., Mastercard ...., or Visa ....

Name ....

Address ....

....

....

Card No. .... Expiration date ....

Signature ....

.... Please indicate the total number of items ordered,

.... total price,

.... tax, if applicable (NY and CA residents)

.... total enclosed

ORDER FORM (Outside North America)

Order from your bookstore or mail to
Customer Services Department, Cambridge University Press, Edinburgh Building,
Shaftesbury Road, Cambridge CB2 2RU, U.K.

.... 43064-X   Numerical Recipes in FORTRAN: The Art of Scientific Computing,
Second Edition   £35.00

.... 43717-2   FORTRAN Diskette (IBM, 5¼ inch/1.2M), second edition   £24.95

.... 43719-9   FORTRAN Diskette (IBM, 3½ inch/720K), second edition   £24.95

.... 43716-4   FORTRAN Diskette (Macintosh), second edition   £24.95

.... 43721-0   FORTRAN Example Book, second edition   £19.95

.... 43108-5   Numerical Recipes in C: The Art of Scientific Computing, Second Edition
£35.00

.... 43714-8   C Diskette (IBM, 5¼ inch/1.2M), second edition   £24.95

.... 43724-5   C Diskette (IBM, 3½ inch/720K), second edition   £24.95

.... 43715-6   C Diskette (Macintosh), second edition   £24.95

.... 43720-2   C Example Book, second edition   £19.95

The following first edition products are still available:

.... 37516-9   Numerical Recipes in Pascal: The Art of Scientific Computing   £30.00

.... 37532-0   Pascal Diskette (IBM)   £21.50

.... 38766-3   Pascal Diskette (Macintosh)   £26.50

.... 37675-0   Pascal Example Book   £19.50

.... 37533-9   Pascal Example Diskette (IBM)   £21.50

.... 38767-1   Pascal Example Diskette (Macintosh)   £26.50

.... 40689-7   Numerical Recipes Routines and Examples in BASIC   £19.50

.... 40688-9   BASIC Diskette (IBM)   £21.50

Name .... (Block capitals please)

Address ....

....

....

Please accept my payment by cheque or money order in pounds sterling:
I enclose (circle one) a Cheque (made payable to Cambridge University Press)/UK Postal
Order/International Money Order/Bank Draft/Post Office Giro.

Please accept my payment by credit card:
Charge my (circle one) Barclaycard/VISA/Eurocard/Access/Mastercard/Bank Americard/
any other credit card bearing the Interbank symbol (please specify).

Card No. .... Expiry date: ....

Signature .... Date: ....

Address as registered by card company: ....

....

....

Prices of diskettes do not include V.A.T., which should be added to all U.K. purchases.